CREATIVE DESIGN ENGINEERING

CREATIVE DESIGN ENGINEERING
Introduction to an Interdisciplinary Approach

TOSHIHARU TAURA
Department of Mechanical Engineering
Faculty of Engineering, Kobe University

Amsterdam • Boston • Heidelberg • London
New York • Oxford • Paris • San Diego
San Francisco • Singapore • Sydney • Tokyo

Academic Press is an imprint of Elsevier

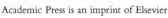

Academic Press is an imprint of Elsevier
125 London Wall, London EC2Y 5AS, UK
525 B Street, Suite 1800, San Diego, CA 92101-4495, USA
50 Hampshire Street, 5th Floor, Cambridge, MA 02139, USA
The Boulevard, Langford Lane, Kidlington, Oxford OX5 1GB, UK

Notices
Knowledge and best practice in this field are constantly changing. As new research and experience broaden our understanding, changes in research methods, professional practices, or medical treatment may become necessary.

Practitioners and researchers must always rely on their own experience and knowledge in evaluating and using any information, methods, compounds, or experiments described herein. In using such information or methods they should be mindful of their own safety and the safety of others, including parties for whom they have a professional responsibility.

To the fullest extent of the law, neither the Publisher nor the authors, contributors, or editors, assume any liability for any injury and/or damage to persons or property as a matter of products liability, negligence or otherwise, or from any use or operation of any methods, products, instructions, or ideas contained in the material herein.

Library of Congress Cataloging-in-Publication Data
A catalog record for this book is available from the Library of Congress

British Library Cataloguing-in-Publication Data
A catalogue record for this book is available from the British Library

ISBN: 978-0-12-804226-7

For information on all Academic Press publications
visit our website at https://www.elsevier.com/

Working together
to grow libraries in
developing countries

www.elsevier.com • www.bookaid.org

TABLE OF CONTENTS

LIST OF FIGURES

LIST OF TABLES

PREFACE

Traditionally, engineering has primarily been concerned with *how* to make things. However, as technology has advanced, and we have no shortage of things, a new challenge for today's engineers is *what* to make. In tackling this, design has come under the spotlight as a keyword. The kind of design that hopes have been pinned on, however, is not about adding stylistic or artist value. Rather, it refers to the synthetic creation of innovative artifacts in line with social feelings and criteria, or awareness of problems. This book provides an academic introduction to the theory and methodology of *innovative creation through design in the broad sense: design to meet the desires of contemporary society* (this is called **creative design engineering** in this book).

This book is informed by the following ideas. First, design is considered a kind of universal human act. Second, an interdisciplinary approach that brings together perspectives from fields such as cognitive science and knowledge science is adopted. Third, the scope of the discussion also includes the process of creating an initial idea for a new product (this will be called the pre-design phase), as well as the use of the product in society (the post-design phase).

We will develop the discussion by examining the fundamental questions of *why we design* and *why we are able to design*. We need to return to basic questions like these if we want to safely and constantly provide society with products that are highly novel and contain the potential to be highly risky—those that absolutely must not fail. A hypothesis to answer the question of *why* is also necessary to engage in a systematic and scientific discussion of design.

This book also presents a specific methodology for innovative creation through design in the broad sense. In particular, in addition to a systematic discussion of concept generation methods, we will present an overview of design methodology relating to the early stage of engineering design (conceptual design) based on Pahl and Beitz's internationally acclaimed *Engineering Design—A Systematic Approach* (Springer).

We expect to see *creative design engineering* being increasingly incorporated into undergraduate and graduate courses of study. I would like to see this book used as a textbook to accompany such classes. I have therefore tried to make this book as easy to understand as possible. Nevertheless, I have included several recent research findings as the topic is still very new, even in academic circles. That said, the book's presumed purpose as a textbook is merely to guide my writing. Actually, I would be delighted to see it studied by engineers or design researchers, as well as read by a wider, non-specialist audience.

This book has been translated from the original Japanese book titled *Sōzō Dezain Kōgaku* (Creative Design Engineering) (published by the University of Tokyo Press, 2014).

Toshiharu Taura
December 2015

CHAPTER 1

Purpose of Creative Design Engineering

Contents

1.1 SCIENCE AND TECHNOLOGY, PRODUCTS, AND SOCIETY

The word "design" is used a lot these days. You will hear it in terms such as industrial design and mechanical design, but also in terms such as corporate design or career design. This is probably because it contains something that is both essential and attractive for society. We place high value on people with synthetic and creative abilities—perhaps because we hope that they will help society overcome its pervasive sense of stagnation. This phenomenon is particularly evident in the *monozukuri* (manufacturing) sector. Until recently, its main role was to consider *how* to make things. However, now that we have no shortage of things, and technology has reached fruition, *what* to make has become the new task. Meanwhile, environmental problems have escalated, and there is a growing awareness of the need to restructure the very systems that society is founded on. These issues are large scale and complex. Design seems to have been identified as a keyword in addressing them. Accordingly, there is an increasing need to understand design from a general and meta level, as well as from a practical perspective.

Creative Design Engineering
http://dx.doi.org/10.1016/B978-0-12-804226-7.00001-6

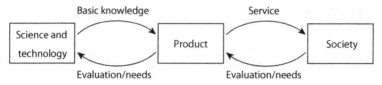

Figure 1.1 The relationship among science and technology, products, and society.

Engineering design and industrial design are closely related to science and technology,[1] with the goods produced in these areas of design often incorporating the most cutting-edge technology. Examples include airplanes, cars, and computers. A discussion of present and future design is not possible without addressing science and technology.

Goods, on the other hand, are more closely related to society.[2] They are used and disposed of within society, and this is where new needs are also generated. We could rephrase this as follows: goods that nobody uses are meaningless. Their value is directly dependent on whether they are accepted by society. Along with goods such as cars and consumer electronics, this book also includes power stations and other industrial plants under the category of objects of design. From here onward, we will therefore refer to all objects of design as **products**.

The relation among science and technology, products, and society is summarized in Fig. 1.1.

Science and technology enables the development of products, the desired function of which can be obtained with fundamental knowledge about physical phenomena. At the same time, products also require science and technology to discover new principles and improve on existing ones. For example, internal combustion engines are designed based on knowledge of thermodynamics and strength of materials. Science and technology is then called upon, for example, to develop more innovative ways of powering a car, such as fuel cells.

Products, in turn, provide society with a service (utility or a sense of satisfaction), and they also receive from society an evaluation of these services and feedback on new needs. For example, gasoline-powered cars provide

---

[1] The term "science and technology" is used as a generic term for an understanding of the basic principles of physical phenomena and the basic knowledge required to apply it. Science and technology is also considered an element of design. The relation between science and technology and design will be discussed in greater detail in Chapter 12.

[2] In this book, the term "society" is used to represent the world inhabited by people where they use products rather than a specific community.

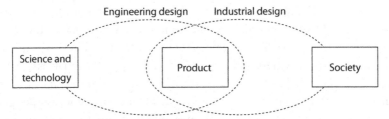

Figure 1.2 Engineering design and industrial design.

users with the convenience of transportation, while they receive demands from users for new functions, such as high fuel efficiency.

The relations shown in Fig. 1.1 also help us to understand the difference between engineering design and industrial design. Whereas engineering design is closer to science and technology, in the sense that it relies on a fundamental knowledge of physical phenomena, industrial design is closer to society, in the sense that the focus is on relations with consumers (Fig. 1.2).

1.2 PERSPECTIVES ON DESIGN: ANALYTICAL AND SYNTHETIC METHODS

This book uses the term "design" in the broad sense, encompassing engineering design and industrial design. The word "design" can be used in different ways. For example, "I like the design of that chair," "This event is well designed," or "Let's design a vision for the company." In each of these sentences, "design" expresses a different meaning. In the first, it refers to an esthetic appreciation, in the sense of giving color and shape to products. In the second, it refers to structure and order, in the sense of orchestrating a plan to achieve a goal. In the third, it refers to drawing a vision for the future, in the sense of pursuing an ideal image. In this book, we will focus on the second and third meanings.

Design methods can largely be classified into two types.

The first method is primarily used *when there is an explicit problem*. A **problem** usually refers to the gap between an existing situation and a goal. In this book, we use it to refer to such a gap in relation to *a specific product or how it is used*. The key here lies in *analyzing* the goal and the existing situation. In this book, the term **analysis** means understanding the nature of something *that already exists* by breaking it down into several parts or constituent characteristics. Accordingly, we refer to the design method that focuses on analyzing a problem as the **analytical method**. Here, the purpose of design is to solve a problem. In this case, the reason for the design (this will be called

the **motive of design**—discussed in further detail in Chapters 2 and 3)
lies in the problematic product or situation that creates a need for it. For
example, the motive for designing the washing machine was to make it easier
to do the laundry, which until then had been a laborious manual task. Subsequent improvements have since been made in response to demands for better machines. Here then, the reason why the washing machine was designed
(the motive) lies in the laundry situation (the task of laundry being arduous),
and later, the washing machine itself (defects in the machine and scope for
improvement). Similarly, the motive for designing the bridge arose out of the
need to cross rivers. Accordingly, the washing machine and the bridge were
both designed to solve problems: the laborious task of hand-washing and imperfect washing machines in the case of the washing machine, and difficulty
crossing rivers and imperfect bridges in the case of the bridge.

The second method is primarily used *when there is no explicit problem*.
Products are still designed even if no problem has been defined. This is
where designers pursue their inherent **ideals**. With this kind of design, the
resulting product and how it will be used are of course not specified since
there is no explicit problem. The focus of the design, therefore, is shifting
from an analysis of existing things to a new creation (**synthesis**). Synthesis,
as a notion, is the opposite of analysis. Conventionally, it is used in the sense
of combining various things *that already exist* into something *that does not
yet exist*. Still, synthesis alone is not enough to constitute the kind of design
that pursues an ideal. This is because combining things that already exist also
requires that they are analyzed. The method that encompasses both analysis and conventional synthesis is referred to in this book as the **synthetic
method**. In this method, the direction of the design is not uniquely determined. This means that the design is often developed based on the feelings
and criteria that exist within the designer's mind. This goes beyond gut or
irrational feelings; rather, it involves the designer carefully considering what
would be ideal at that particular time while drawing from the standards in
his or her mind. This could also be described as design based on an autonomous motive. This means that the motive for the design comes from within
the designer himself or herself.

It should be noted that the synthetic method sometimes also applies to
situations where there is an explicit problem. For example, a space shuttle is
understood as a combination (or synthesis) of a rocket and an airplane (this
will be discussed in further detail in Chapter 5). Conversely, even if there
is no specific problem, a design can draw on existing things without the
occurrence of synthesis. Thus, the analytical method alone is not necessarily

always used where there is an explicit problem, and the synthetic method alone is not necessarily always used where there is no explicit problem.

These categories correspond to the classification of motivation into extrinsic and intrinsic (Ryan and Deci, 2000). *Extrinsic motivation* refers to instances of motivation where an external reward is received. *Intrinsic motivation* is driven by curiosity. Strictly speaking, motive and motivation are not the same: motive refers to the reason, while motivation refers to the will. A design method involving the motive of solving a problem inherent to a product or its situation is largely related to extrinsic motivation, while design involving the motive of giving expression to the designer's feelings and criteria is largely related to intrinsic motivation.

To further develop the discussion, we could define **design** as "the process of composing what is desirable toward the future" (Taura and Nagai, 2012). Let us examine this definition more closely.

The phrase "toward the future" specifies a temporal element for design. The notion of "future" is extremely abstract. For example, we can't draw a picture of the notion of the future. It is possible to picture something—the city where you live or your life, perhaps—at some point in the future, but not the future itself. The actual notion of the future can only be understood and expressed through *language*. In the context of design, the future is thought to have two meanings. First, it is seen as something that can be grasped inductively, such as a marketing forecast. Second, it is the desire or ability to perceive or express driven by an inner trigger, such as the mind's eye of an artist. We could argue that, in design, it is necessary to look to the future in both of these senses.

The phrase "what is desirable" determines the object of the design. This would include that which can be clearly derived from existing problems, as well as the ideal future form that is desirable to pursue. In the case of the latter, this would involve the designer closely examining what is ideal while drawing from his or her inner standards. One such standard is *a sense of resonance*. With man-made objects, *naturalness* is a scale for evaluating what is desirable. However, simply approximating something that is found *in the natural world* is not necessarily enough to create something that resonates with our hearts. In fact, some things that do not exist in the natural world nevertheless resonate deeply with our hearts. Music is an example of this. Most man-made music is very different from sounds that exist in the natural world. Yet, it deeply resonates with us. Therefore, in order to approach the desirable figure of a man-made object, rather than imitating something in the natural world, the designer needs to search the depths of his or her heart for a sense of resonance.

"Composing" refers to the design process. Design often involves creating things that do not exist in the natural world. This involves, as discussed earlier, an analysis of the current situation and conventional synthesis. Here, accordingly, the term "compose" corresponds to the notion of "synthetic method" that was discussed earlier.

Next, we need to define **creativity** as it applies to design. Building on our definition of design, creativity in design could be considered as an evaluation of whether or not the design is creative, with the standards being the extent to which it has approached what is desirable. Thus, it could be defined as a measure of proximity to what is desirable. In other words, novelty—which is generally used in conjunction with creativity—is not a causal factor of creativity, but rather a result. It therefore follows that pursuing something merely on account of its uniqueness can never approach what is desirable.

1.3 THE MAIN QUESTIONS: WHY DO WE DESIGN AND WHY ARE WE ABLE TO DESIGN?

Let us consider the idea of **creative design engineering**, which is also the title of this book. Rather than being a simple amalgamation of engineering design and industrial design, this title seeks to position design as a universal human act that encompasses science and technology, and society. The term "creative design" encompasses seeking what is desirable, and also includes "creative thinking." "Engineering" recognizes that a discussion of present and future design needs to include science and technology, and it also incorporates a sense of looking beyond abstract theory and toward a practicable methodology.

The term **concept** is often used in the field of design. In engineering design, the early stage of design is referred to as **conceptual design**. This is when thought is given to the basic structure and layout of the product. The term "concept" can also be used to refer to ideas, but in this book—in the context of design—it is defined as the figure of an object, along with other representations such as attributes or functions of the object, which existed, currently exist, or might exist in the human mind as well as in the real world (Taura and Nagai, 2012). I would like to address the questions, Why do we design? and Why are we able to design? by considering the process of generating and manipulating a concept, the subsequent process of actualizing the product based on that concept, and the process by which the science and technology gains acceptance in society.

Let us look at the first question, Why do we design? On a basic level, most individual engineers or designers design because this is what they are employed to do and/or in response to client requests. In fact, you may even be wondering what the point of this question is. However, we need to go back to the fundamental question of why people design if we are going to take the lead in actualizing designs that will gain acceptance in the real world. This question does not concern the kind of contracts under which individual engineers or designers carry out their work, or what specific needs society has. Rather, it is concerned with the reasons behind the existence of products if we look at design within society as a whole. For example, let us consider the car. Each year, a large number of cars are designed and manufactured. While cars offer convenience, they are also dangerous and could potentially cause a lot of accidents. Today, very few people would question whether we should actually be making cars. However, at some point in the past, did this discussion occur? If we assume not, this would mean that we have been making cars without really knowing the reason (motive) of design. The question as to why we should design cars, to date, has not been discussed; it has not been considered necessary.

On the other hand, developing novel products that transcend traditional product categories has become an important challenge in today's world of plenty: designers need to pioneer the creation of products that anticipate society's needs. We therefore need to look at the source of such products, which lies behind these needs—rather than simply pursuing superficial or explicit needs.

This question has not been given much conscious thought; it is unlikely that this has even been deemed necessary. However, particularly with advances in science and technology and the large-scale and complex nature of contemporary issues, it is time for a conscious and rigorous discussion. In this book, we consider the motive of design to arise out of *(inner) feelings and criteria, or awareness of (outer) problems* relating to a product and its use (discussed in further detail in Chapter 2). We will discuss the nature of motive of design and discuss how to anticipate it. Specifically, we will try to identify and describe the motive of design by means of a rational and experimental examination.

The second question is why we are able to design. Again, this question may initially strike you as vague and meaningless. Indeed, we can design without thinking about why we are able to. However, this question plays an important role in the systematic and scientific study and practice of design. Systematic and scientific methods rely on the repeated testing of hypotheses relating to why a phenomenon arises. Without a hypothesis to explain why people are able to design, we will not be able to identify a systematic and scientific method for the study and practice of design. Nor

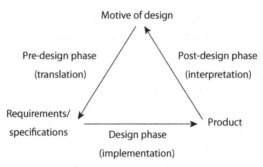

Figure 1.3 The design cycle model.

will we be able to discuss the methodology associated with design. If we are going to treat design as a subject of academic inquiry, we need to directly address this question.

1.4 THE DESIGN CYCLE MODEL: PRE-DESIGN PHASE, DESIGN PHASE, AND POST-DESIGN PHASE

The design process—in the narrow sense—can be defined as the process of developing a design solution to fulfill external requirements or specifications[3] (discussed in further detail in Chapter 9). In this book, we expand on this definition by looking at design as a cycle (Taura, 2014). We need a broader understanding of the design process before we can discuss the two questions mentioned in the previous section. Our model (referred to here as the **design cycle model**) consists of three phases: the **pre–design phase**, the **design phase**, and the **post–design phase**. This will be discussed in further detail in Chapter 2, but for now, the overview in Fig. 1.3 provides a brief introduction.

The pre–design phase refers to the phase when the concrete requirements or specifications for a new product that we might expect society to accept are generated on the basis of the motive of design. It is a "translation" process, during which the motive of design is translated into the requirements or specifications for the new product.

The design phase refers to the conventional design process. During this phase, structures and shapes are developed to satisfy the requirements or specifications for the new product. It could also be described as a process of "implementation," during which the requirements or specifications are implemented into a specific form. Usually, the design phase refers only to

---

[3] The term "specifications" is used here in the general sense. Its usage in mechanical design is explained in Section 9.1.

the process of creating drawings or sketches. Here, however, we also include the process of manufacturing the actual product.

The post-design phase refers to the generation—explicitly or implicitly, and in society or individuals—of the motive of design, which will lead to future products. Not only do consumers use the product in accordance with its pre-defined usage, they also develop an understanding of the product and discover new uses or meanings for it. This phase could therefore be described as a process of "interpretation."

The term "design," as defined in Section 1.2 of this chapter, encompasses the pre-design and design phases of the design cycle model. The post-design phase is not a direct part of design, although it is closely related.

1.5 REQUISITE KNOWLEDGE

Designing requires knowledge of various disciplines. For example, to design a car engine, you would need to know about thermodynamics, hydrodynamics, and strength of materials, as well as mechanical components. To design the body of a car, you would need to know about hydrodynamics and vibration engineering, as well as form and esthetics. However, while various kinds of knowledge are required in order to design, knowledge alone is not enough to be able to design well. Most of the aforementioned knowledge is related to the structure and shape of the car. It enables the designer to analytically infer the characteristics and behavior of *existing things*. However, design involves creating something *that does not yet exist*. The designer, therefore, first needs to posit the structure and shape in some way.

For example, imagine that you were asked to design and build a tall, robust, and esthetically pleasing tower using 13 strands of spaghetti and some tape. Knowing about the characteristics of spaghetti and being good at using tape would help you create a good structure and shape. However, this does not necessarily mean that you will be able to come up with a good design. In other words, an ordinary tower can be made without special knowledge or experience.

However, what is important here is producing an *outstanding* design to fulfill a request. What would you need to know and what would you need to be able to do in order to produce an outstanding design? This is actually very difficult to answer. The ability[4] to design at an outstanding level

---

[4] In this book, the notion of ability or capability and related methods are discussed in general terms, without taking into account how they apply to individuals, groups, or organizations.

(**design capability**[5]) is not found in a formula, and there are no textbooks setting out the solution.

It is undeniable that some designs are better than others. For example, if you were actually to attempt to design and build the spaghetti tower, you would encounter a wide range of different ideas. As discussed, no special knowledge or experience is needed. This can be compared to trying to develop a theory and methodology to understand how a world-class sprinter runs, though we know that everyone can run.

When studying design, students carry out exercises to develop their design capability. After all, it is important to merely attempt something. Similar to how writing exercises are required to improve writing skills, it is essential to make attempts at designing things. That said, if there was even something vaguely resembling a theory and principle of design capability, we would be able to use an understanding of this to deliberately study and practice design. If, hypothetically, we knew that there was a kind of cognitive framework at work behind outstanding design, we would then be able to strengthen it so as to systematically and scientifically study and practice design.

1.6 STRUCTURE OF THIS BOOK

The remainder of this book is structured as follows. Part 1 (Chapters 2 and 3) discusses the motive of design. Part 2 (Chapters 4–8) constitutes a systematic examination of concept generation at the pre-design phase. Part 3 (Chapters 9 and 10) provides an overview and analysis of conceptual design during the design phase. In Part 4 (Chapters 11 and 12), we discuss the competence necessary to execute the design process and the methods to effectively acquire and transfer such competence. We also look at the post-design phase and discuss the process by which the science and technology and the associated product gain acceptance in society. Finally, Chapter 13 concludes the discussion, covering the nature of design and the mindset of designers in the future. We also seek to answer the key questions, Why do we design? and Why are we able to design?

Please note that in this book, **concept generation** refers to the process of coming up with and expanding on ideas to realize the requirements or specifications for the new product. This occurs at the final stage of the pre-design phase and the very early stage of the design phase. Meanwhile, **conceptual design** refers to the early stage of the design phase.

[5] See Section 13.2 for more on design capability.

Task 1[6]

1. In groups of 4 or 5, design a tower that can be built using 13 strands of spaghetti and some tape. The evaluation criteria are height, robustness, and beauty. (15 min)
2. Build the tower. (15 min)

Task 2

Identify what you need to learn in order to be able to design an outstanding spaghetti tower.

REFERENCES

Ryan, R., Deci, E., 2000. Self-determination theory and the facilitation of intrinsic motivation, social development, and well-being. Am. Psychol. 55 (1), 68–78.

Taura, T., 2014. Motive of design: roles of pre- and post-design in highly advanced products. In: Chakrabarti, A., Blessing, L. (Eds.), An Anthology of Theories and Models of Design. Springer, London, pp. 83–98.

Taura, T., Nagai, Y., 2012. Concept Generation for Design Creativity – A Systematized Theory and Methodology. Springer, London, pp. 9–20.

---

[6] This task is based on an activity carried out at a workshop in Seoul in 2007 organized by Yong Se Kim.

PART 1

Motive of Design and the Design Cycle Model

CHAPTER 2

Motive of Design and the Design Cycle Model

Contents

2.1 SOCIAL MOTIVE AND PERSONAL MOTIVE

In this book, the reason for designing something is called motive of design (Taura, 2014). In most cases, human action is underpinned by some reason. Such reason may be implicit or explicit. It may also be unconscious. Design supplies products to society, and products accordingly have some kind of effect on society. Just as that effect can enrich our lives, it can also cause accidents or damage. When this occurs, it is not enough to simply claim that the product was unconsciously designed.

Motive of design does not refer to the need for products, or requirements or specifications for products. Nor is it the compensation received by individual engineers or designers from organizations, or those requesting the design. Rather, it is the reason why need arises, why this need forms requirements or specifications of products, and why compensation is paid to engineers and designers. When cars are made, the requirements or specifications for design are determined, such as the amount of fuel required for a certain vehicle mass to attain an expected comfort and safety level. In a narrow sense of the word, to design is to derive a specified structure and shape of the product from such requirements or specifications in order to satisfy them.

We must ponder over how these requirements or specifications are determined. Why do we make cars in the first place? They may be convenient, but at the same time, cars are dangerous objects that cause many accidents.

Creative Design Engineering
http://dx.doi.org/10.1016/B978-0-12-804226-7.00002-8

So how do we draw the line between danger and convenience? We are able to estimate the likelihood of accidents occurring and even how well a vehicle can protect the passengers when an accident occurs. However, do we possess the procedure for determining the level of danger that society is willing to accept from cars? It is assumed that this is decided through actual experiences. Cars are driven, accidents occur, society reflects upon these, and somewhere within the repetition of this cycle, the product is (or appears to be) accepted by society. To put it differently, perhaps the acceptance of cars within society came first, and the reason for such acceptance was not a topic of prior debate. Although too late to discuss, the reason why we began designing these convenient but dangerous cars may exist in an implicit and intrinsic societal value.

In 2011, the cofounder of Apple Inc., Steve Jobs, passed away. He has been quoted as saying, "we try to use the talents we do have to express our deep feelings" (Isaacson, 2011). Generally, product development is sensitive to the opinions of users. The normal practice is to survey customer needs in order to create a product that meets those needs. We might say that Jobs, however, found the reason for pursuing the development of products within his own mind.

It is not the needs, requirements or specifications, or compensation, but rather the reasons behind these and from which these are derived, that this book terms the motive of design. As stated in Chapter 1, the motive of design has rarely been explicitly discussed. However, while the creation of innovative products is required in the present and the foreseeable future, highly advanced technologies and the products that incorporate them have the potential to bring about accidents that could cause irreparable damage. This means that a detailed consideration of the notion of the motive of design is necessary. As mentioned in Section 1.2, there are two types of motives of design: those motives that arise from the problems that exist in products or the situations that require them, and those that arise from the feelings and criteria that exist within the designers themselves.

Based on the previous argument, this book defines **motive of design** as "the reason behind the design process, arising from 'feelings and criteria,' or 'awareness of problems' relating to a product." Here, "feelings and criteria" are related to motive that exists within the designers, and "awareness of problems" is related to motive that exists within the product and situation. This will be explained further in the next section.

Next, the axes for categorizing motive of design will be examined. The first is whether or not the feelings and criteria, or awareness of problems are shared in society. Although the term motive is used with regard to individuals generally, in the design of products, the feelings and criteria, or awareness of problems that are (or are thought to be) shared in society, either explicitly or implicitly, play an important role. Thus, this book will refer to motive held by individuals as **personal motive** and that (regarded as being) shared in society as **social motive**.

2.2 OUTER MOTIVE AND INNER MOTIVE

In addition to social motive and personal motive, motive can be categorized using one more axis as suggested in Sections 1.2 and 2.1. That is, motive can be classified based on whether it externally exists within the product and the situations that require that product, or internally within the designer's mind. This book will refer to the former as **outer motive** and the latter as **inner motive**.

Outer motive is related to the first method of design mentioned in Section 1.2, whereby the basis for design is individual or social problems. This can be easily inferred from the definition in Chapter 1 of "problems" as gaps between the existing situation and goal of a particular product or how it is used. In actuality, many of the design process models proposed in the past consider this process within the framework of problem solving. In problem-solving scenarios, the goal is often clear; examples include remedies for natural disasters, accidents and other adverse situations, or responses to customer demands. In such cases, the problem-solving style of design usually consists of an analysis of the nature of the gap between the goal and the existing situation, and the formation of a problem-solving strategy based on this analysis, as outlined in Chapter 1. Thus, the outer motive can be said to be related to the analytical method, which drives design aimed at solving existing problems.

Inner motive, on the other hand, relates to design not for solving existing problems, but for pursuing ideals. In this category of design, importance is placed on what the "ideal" should be. In engineering design, this refers to the devising of the ideal functions with which future man-made objects should be equipped, and in industrial design, it refers to devising shapes and interfaces that create an ideal impression on users. It can be assumed that the designer's inner feelings and criteria play an important role in determining this ideal. This is because what is ideal to the designer is actually based on standards within her/his own mind. Based on these standards, the designer then synthesizes new concepts with reference to existing ones. Therefore, it

Table 2.1 Categories of motive of design

	Outer motive	Inner motive
Personal motive	Personal outer motive	Personal inner motive
Social motive	Social outer motive	Social inner motive

can be assumed that inner motive is related to the synthetic method, used to generate ideal concepts that do not yet exist.

The categorization of the motive of design from the previously mentioned two axes is summarized in Table 2.1. Here, **social outer motive** is the motive arising from awareness of problems, which is shared by members of society, such as "we need a bridge to cross that river." **Personal outer motive** is the motive arising from individual awareness of problems, which is not shared in society, such as "I want shoes that fit the shape of my foot." **Social inner motive** arises from feelings and criteria that are (believed to be) shared in society, such as "we all want to travel to see a beautiful sunset." Finally, **personal inner motive** arises from individual feelings and criteria that are not (believed to be) shared in society, and are related to feelings and criteria similar to the "sense" possessed by Jobs. Discussion in this book will focus on social outer motive and personal inner motive. This is because it is believed that outer motive tends to be easily shared in society while inner motive tends to be contained within individuals. Hereafter, social outer motive will be referred to simply as "outer motive," and "inner motive" will be used to refer to personal inner motive.

2.3 THE DESIGN CYCLE MODEL AND MISSING LINKS

The design cycle model, comprised of the pre-design phase, design phase, and post-design phase, was outlined in Section 1.4. This section examines further the two main questions raised by this book, by focusing on the missing links (noncontinuity) in each phase. This is because, in addition to design sometimes being carried out without fully developed social motive in the pre-design phase, social motive is sometimes formed (meaning that implicit needs are generated resulting in social acceptance of a product and the subsequent development of a new product) without interaction between the user and the product (ie, experience of utility or accident) during the post-design phase. We can presume that such missing links are related to the essence of our two main questions. Specifically, phases can be classified into three categories: *continuous, semi-continuous*, and *non-continuous*.[1] The term **continuous** is used to mean that *an explicit basis exists* in the phase for almost invariably producing the same result. **Semi-continuous** is used to mean that the result of the phase *has an implicit basis*, although this is not fixed (results may differ depending on who conducts the phase). The term **non-continuous** is used where the result of the phase *has neither an implicit nor explicit basis*.

2.3.1 Categories of the Pre-design Phase

The pre-design phase generates requirements or specifications for new products. This book will focus on the motive of design as the basis for generating these requirements or specifications. It will also categorize the pre-design phase based on whether the motive of design is social motive or personal motive, and whether it is implicit or explicit.

2.3.1.1 Continuous Pre-design Phase

Continuous pre-design phase is a process whereby *social motive* is created and contained *explicitly* in society, and the requirements or specifications for new products are generated from that social motive. An example of this would be the improvement of utility or efficiency on the basis of data obtained through the usage of products.

2.3.1.2 Semi-continuous Pre-design Phase

Semi-continuous pre-design phase is a process whereby *social motive* is created and contained *implicitly* in society, and the requirements or specifications for new products are generated from that social motive. Specifically, requirements or specifications for new products are determined through methods such as market surveys, antenna shops, and observations of user behaviors.

2.3.1.3 Non-continuous Pre-design Phase

Non-continuous pre-design phase is a process whereby *social motive* is neither implicitly nor explicitly created, and the requirements or specifications for new products are generated from the *personal (inner) motive* of the designer. This includes, for example, the determination of requirements or specifications for completely new and highly innovative products. There are two ways to progress through non-continuous pre-design phases. One is

---

[1] In Taura (2014), these phase categories are referred to as "deductive," "inductive," and "abductive." However, since Chapter 7 of this book uses the terms "deduction," "induction," and "abduction" to explain a method for inferring, design phases will be described in this book as "continuous," "semi-continuous," and "non-continuous" to avoid confusion.

to hypothesize the kinds of feelings and criteria, or awareness of problems, that would cause society to accept the product. The other is to pay little heed to society and create the new product based on one's own ideas. In this case, the requirements or specifications of the product risk becoming self-serving, and consequently being in disagreement with the feelings and criteria, or awareness of problems that are shared in society. On the other hand, if products designed in this way were to be accepted by society, the products would be pre-empting the social motive of design.

2.3.2 Categories of the Design Phase

The design phase results in the creation of a new product. This phase refers to design in the narrow sense, and in engineering design, it corresponds to the design process. This book will focus on the fact that the characteristics of most products such as their structure and shape are devised by basing them on existing products. The design process will be categorized based on the degree to which this occurs (Pahl and Beitz, 1995).

With regard to the originality of the design process, Pahl and Beitz (1995) propose the following classifications.

2.3.2.1 Original Design

This involves the creation of an original **solution principle** for a system (plant, machine, or assembly) for the same, a similar, or a new task (finding a solution principle and the method for combining these are explained in detail in Chapter 9).

2.3.2.2 Adaptive Design

This involves adapting a known system to a task with a different objective, maintaining the same solution principle. This often involves original design of parts or assemblies.

2.3.2.3 Variant Design

This involves varying the size and/or arrangement of certain aspects of the chosen system without making any changes to the functions or solution principles. In this type of design, changes in materials, constraints, or technological factors do not give rise to any new problems.

Original design involves the creation of an *original* solution principle, and therefore is non-continuous. Adaptive design involves *transferring* the use of an existing solution principle to a product with a different objective, and therefore is semi-continuous. Variant design *does not alter* the solution principle, and therefore is continuous.

2.3.3 Categories of the Post-design Phase

The post-design phase results in the creation of new motive of design that leads to the next product. This book will focus on the interaction between users and products (ie, experience of utility and accident) that forms the basis for the creation of motive of design. The post-design phase will be categorized based on whether the interaction is direct, indirect, or has yet to be experienced, as well as whether the product was used in the way intended by the designer.

2.3.3.1 Continuous Post-design Phase

Continuous post-design phase refers to a process whereby the feelings and criteria, or awareness of problems surrounding products, come to be shared in society through *direct interactions* between users and products, *without going against the usage intended by the designer*, which leads to the design of the next product. An example of this would be if the use of a product in line with its guidebook or manual leads to the identification of an aspect that made it less convenient to use, which becomes the motive for improving the product.

2.3.3.2 Semi-continuous Post-design Phase

Semi–continuous post-design phase refers to a process whereby new social motive is created through the *indirect interaction* between users and products (witnessing or hearing about the experiences of others who have used the product and/or encountered accidents using it), *without going against the usage intended by the designer*. For example, this occurs when someone who has not directly used the product, hears or sees the experiences of other users, and feelings and criteria, or awareness of problems surrounding the product come to be shared in society, leading to the development of a new product. Specifically, this could happen when product reviews on clothing or information technology lead to the popularity and perspectives about new products start to spread in society.

2.3.3.3 Non-continuous Post-design Phase

Non-continuous post-design phase refers to a process whereby feelings and criteria, or awareness of problems relating to products come to be shared in society *without interaction* between users and products (ie, with experience of neither direct nor indirect utility or accident). Alternatively, users discover a use *outside the intent* of the designer, and this becomes popular in society. Usually, products equipped with highly advanced technology and with a high level of danger are not permitted to cause accidents, as this would result in huge damage. If no accident is caused (the fact that this is unrealistic will be touched upon in Chapter 13), this means that feelings and criteria, or awareness of problems relating to a product, are formed without actual first-hand experience of the product dangers by users. The following situation may occur, for instance, when a product expert uses charts or statistical methods to plainly explain the risks of the product that have yet to be experienced by society, and this results in the acceptance of the product by society before that product has even been seen. On the other hand, where the product is used in a way, which surpasses what was anticipated by the designer, and word of mouth sees this use become popular, this can form the basis of a new motive of design and lead to the development of a subsequent product. However, the use of the product in unanticipated ways can be dangerous and is sometimes prohibited.

2.3.4 Aspects of the Design Cycle Model

Table 2.2 summarizes the discussion of the pre-design, design, and post-design phases based on the continuous, semi-continuous, and non-continuous categories.

Until now, the pre-design phase has been examined in terms of idea generation, market research, and risk management methods. Discussion of the post-design phase has centered on keywords such as usability, emotional design, and user-centered design. The design phase has been a matter of engineering design and other design methodology. However, each of these has been examined in isolation, and they have not been systematized as one. Further, discussion of the topics associated with the fundamental questions that this book seeks to answer, Why do we design? and Why are we able to design? is required. This book hopes to develop a broad discussion based on the design cycle model, in order to gain a view of the big picture of design and uncover its essence.

2.4 CONTINUOUS AND NON-CONTINUOUS DESIGN CYCLES

In the design cycle, the cycle that is composed of continuous or semi-continuous phases will be called the **continuous design cycle** (Fig. 2.1a). Here, a social motive of design forms the basis for the creation of a new product, and experience of the actual use of this product creates new social motive. Progressing through the cycle requires analysis of each of the following: social motive, requirements or specifications of the new product,

Table 2.2 Categories of the pre-design, design, and post-design phases

	Pre-design phase	Design phase	Post-design phase
Continuous	Social motive of design is created and contained explicitly	Variant design (size and/or arrangement of certain aspects vary but no changes to functions or solution principle)	Direct interaction between users and products, without going against the usage intended by the designer
Semi-continuous	Social motive of design is created and contained implicitly	Adaptive design (solution principle is transferred)	Indirect interaction between users and products, without going against the usage intended by the designer
Non-continuous	Social motive of design is created neither implicitly nor explicitly	Original design (original solution principle is created)	No interaction between users and products, or use of product in a way that surpasses designer's intentions

and the use of the created product. Analyzing the social motive helps reveal the requirements or specifications for the new product, these requirements or specifications enable the actual creation of the product, and analysis of the use of this product creates new motive. Thus, the continuous design cycle requires analysis of *pre-existing* social motive, requirements or specifications, or experience of actual use of a product.

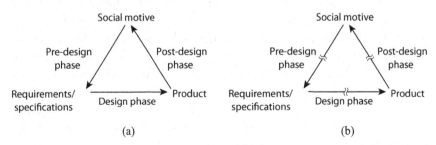

Figure 2.1 (a) The continuous design cycle; and (b) the non-continuous design cycle.

On the other hand, where some or all of the pre-design, design, and post-design phases are non-continuous, this will be called the **non-continuous design cycle** (Fig. 2.1b). Here, the social motive that would create a new product may be undeveloped, no similar products may exist, and the requirements or specifications for generating the product may not be explicitly expressed, or there may be no interaction between the user and the product that would form the basis of the social motive. To progress through the non-continuous design cycle, requirements or specifications must be sought where social motive is not yet fully developed, products must be created where requirements or specifications can only be expressed abstractly, or social motive must be created where the product has yet to be used. This means that this cycle cannot progress through the use of analytical methods alone.

This non-continuous nature of the design cycle makes it more difficult to answer the two main questions of this book, Why do we design? and Why are we able to design? If the cycle is continuous, the answer to the first is that design is carried out based on social motive. If we then ask the question, Why is social motive created? The answer is that interaction between users and products creates social motive. In short, each answer can be found by tracing the design cycle along a clockwise direction. As for the second question, Why are we able to design? the answer is surely our analytical ability. However, if the cycle is non-continuous, then this explanation is not valid. Our attention then should focus on how these missing links come about. In particular, understanding how the missing link proximal to both the pre-design and the post-design phase or at their cross-point in terms of the motive of design is overcome is crucial for answering these two main questions. The key to this lies in the synthetic method outlined in Section 1.2.

The missing link in the pre-design phase is the fact that products are created without the creation of social motive. In this case, as a new product is used, the need for the product becomes more apparent. On the other hand, the missing link in the post-design phase is the agreement in society on whether or not a product will be accepted before the product is actually used. Thus, in the non-continuous design cycle, the social motive (or need) that should precede the product becomes an *afterthought*, and the judgments that should be made after actually using the product and experiencing its utility or safety are made a priori. This means that in the missing link, a *time reversal* occurs. If this time reversal can be overcome, we will see the creation of innovative products that are safe as well as congruent with the needs of society.

Task

Consider what an alien looking down on Earth will see as the reason for human beings making cars. (What do you think is the motive for us to continue making cars?)

REFERENCES

Isaacson, W., 2011. Steve Jobs. Simon & Schuster, New York, pp. 570.

Pahl, G., Beitz, W., 1995. Engineering Design: A Systematic Approach. Springer, Berlin.

Taura, T., 2014. Motive of design: roles of pre- and post-design in highly advanced products. In: Chakrabarti, A., Blessing, L. (Eds.), An Anthology of Theories and Models of Design. Springer, London, pp. 83–98.

CHAPTER 3

Aspects of Motive of Design: One Form of Inner and Outer Motive

Contents

3.1 ONE FORM OF INNER MOTIVE: COMPLEX NETWORKS OF EXPANDING ASSOCIATIONS

What actually is the inner motive of design? If we recall the Steve Jobs quote mentioned in Section 2.1, "we try to use the talents we do have to express our deep feelings," what deep feelings did Jobs have? This section will focus on the human state of mind when forming impressions and generating new concepts, based on the feelings and criteria that comprise inner motive. From this, it will then speculate about the actual characteristics of inner motive. However, deep feelings are not visible. They cannot be directly observed, and thus we must rely on indirect information to hypothesize about them. This section will outline an attempt to do so using simulation research.

3.1.1 Network Structure of the Impression Process

What is in our minds when we form an impression of something we see? For instance, when we see a "lion," we may associate it with "Africa." One by one, our networks of association expand. If we can know what this expansion of associations looks like, we can speculate about how impressions are gained from the object in question. So how can we know this? We could ask experiment participants to say the words that come to their

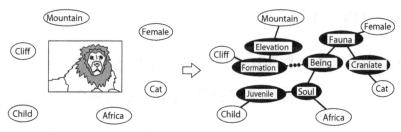

Figure 3.1 Method for generating virtual impression networks.

minds one by one. However, human beings are not able to express all of the impressions they have in their minds; there are also states of mind that cannot be explicitly expressed. Thus, the following method has been tried (Taura et al., 2011). First, participants are asked to describe their impression of an object using a number of words (these were called "explicit impression words"). Next, all (unverbalized) words that could possibly exist (as "paths") between each pair of explicit (verbalized) impression words are searched for. The results of each search are called "inexplicit impression words." All of the paths found are then integrated as one network (this network has been labeled the **virtual impression network**). The network thus generated shows how networks of association expand as impressions are formed. Moreover, if we observe the structure of the network, we can infer the dynamic state of the human mind when forming impressions. Based on this idea, the structure of virtual impression networks was investigated using man-made objects and animals. Fig. 3.1 is a diagram representing the method for generating the virtual impression network.

The words that could exist between each pair of explicit impression words were searched using a semantic network (concept dictionary; specifically, WordNet, Fellbaum, 1998). A semantic network provides the semantic relationships between words, such as hypernym–hyponym and associative relationships. Use of the semantic network allows researchers to search for paths of concepts from one word to another. Fig. 3.2 is a diagram showing path searches using the semantic network. For example, between the impression words "cliff" and "child," inexplicit impression words such as "formation" and "soul" can be found.

Using the aforementioned method, the difference between the expansions of association for "liked" objects and "disliked" objects was investigated. When target objects such as cups and animals were investigated, it was found that the virtual impression networks generated for "liked" objects were more intricately intertwined than those for "disliked" objects. Fig. 3.3 illustrates

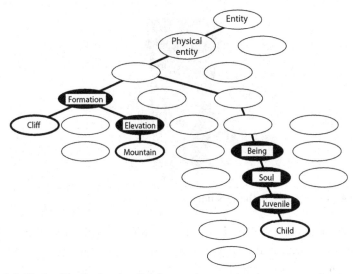

Figure 3.2 Method for finding implicit impression words using a semantic network.

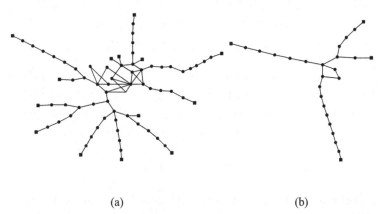

(a)   (b)

Figure 3.3 Comparison between virtual impression networks (a) participant who liked the cup; and (b) participant who disliked the cup. *(Modified from Taura et al., 2011).*

a sample comparison between the expansion of association for participants who answered that they "liked" and "disliked" the cup shown in Fig. 3.4.

The participants who answered that they liked the cup expressed their impressions using the following words (explicit impression words): clarity, glass, ding, ice, liquor, diamond, blueness, antique, costliness, teacup, artisan, polishing, and powder. The participants who answered that they disliked the cup expressed the following impression words (explicit impression words): juice, coldness, glass, and artisan. The virtual impression networks

Figure 3.4 The cup used in the experiment.

created from these explicit impression words using the method outlined previously are indicated in Fig. 3.3. We can see that the virtual impression network of the participant who answered that they "liked" the cup is more complexly structured than that of the participant who answered that they "disliked" the cup. Moreover, a significant difference was almost always found between the two groups of respondents when statistical analysis was conducted using indicators of the complexity and other structural features for all of the networks.

The aforementioned result indicates that good impressions do not follow from a particular concept but rather are strongly related to the way in which associations are created by such impressions. We may say that good impressions *resonate with our hearts*. Does the expansion of associations not have the same effect? Or, we could perhaps assume that the expansion of association is itself a state of "resonance with our hearts."

3.1.2 Network Structure of the Concept Generation Process

Meanwhile, what kind of process is the concept generation process? This process has been simulated using the same method used to infer the impression forming process (Taura et al., 2012). In this simulation, the path between the start point (a pair of concepts expressed using words that will be called **base concepts** in this book; for details, see Chapter 8) and end point (a set of words describing design outcomes) that were obtained from the design experiment (Nagai et al., 2009) was connected virtually. The resulting network will be labeled the **virtual concept generation network**. In the design experiment, participants were instructed to design a new idea by expanding their imagination through combining the pair of words they were given. Specifically, they were provided with pairs of words (base

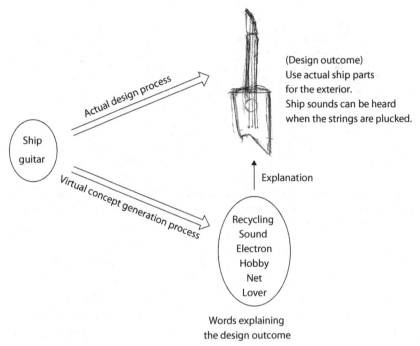

(Design outcome)
Use actual ship parts
for the exterior.
Ship sounds can be heard
when the strings are plucked.

Actual design process

Ship
guitar

Virtual concept generation process

Explanation

Recycling
Sound
Electron
Hobby
Net
Lover

Words explaining
the design outcome

Figure 3.5 Relationship between the virtual concept generation process and actual design process.

concepts)—"ship–guitar" and "desk–elevator." The process of combining pairs of concepts (base concepts) such as these (**two-concept combining process**) is regarded as the simplest and most essential concept generation process. This is examined in detail in Chapters 4–8 of this book. The design outcomes of the design experiment were in the form of sketches and sentences. The simulation did not use the design outcomes themselves as end points, but rather the *sets of words describing design outcomes*. These sets of words were obtained from participants by asking each of them to provide a list of words describing their design outcome. Although these words are not the design outcomes themselves, it is considered that they capture the content of these outcomes. Fig. 3.5 shows the relationship between the virtual concept generation process and actual design process.

The simulation first searches for all the words that could possibly exist between each base concept word and each word describing the design outcome. This search is conducted on every possible combination between base concept words and words describing design outcomes. Then, every path generated by the search is integrated to form one network (the virtual concept generation

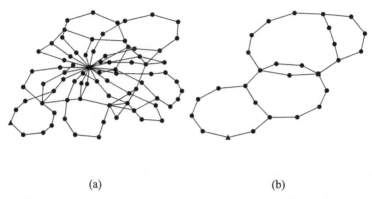

(a)   (b)

Figure 3.6 Comparison between virtual concept generation networks (a) process with more originality in the design outcome; and (b) process with less originality in the design outcome. *(Modified from Taura et al., 2012).*

network). As in the impression simulation, searches for possible words are performed using the semantic network (WordNet, Fellbaum, 1998).

The study then investigated the difference in structures between the virtual concept generation networks generated by the set of words describing the design outcomes with more originality and the design outcomes with less originality. The design outcomes with more originality had more intricately intertwined networks than did those with less originality. Fig. 3.6 exhibits a sample of virtual concept generation networks generated from two designs with the base concepts "ship–guitar." The words used to describe the design outcome with more originality were recycling, sound, electron, hobby, net, and lover. Meanwhile, those used to describe the design outcome with less originality were alarm clock and novelty. Fig. 3.6 illustrates the virtual concept generation networks obtained from these words. We can clearly see that the network for the outcome with more originality is more complex compared to that of the outcome with less originality. Furthermore, when statistical analysis was conducted on all of the design outcomes obtained through the design experiment using indicators of structural features of the networks such as complexity, a significant relationship was almost always found between the two.

This result indicates that good ideas are not born out of particular concepts but rather are strongly related to the associations relevant to the process of generating that concept. It may be said that one's *heart dances* when one has a good idea. Does the expansion of associations not have the same effect? Or we could perhaps assume that the expansion of association is itself a state of "heart dancing."

3.1.3 The Association Process in Overcoming the Missing Links in Inner Motive

The results of the two simulations illustrated the intricately intertwined and expanding associations involved in the process of forming a good impression and of coming up with a good idea. Furthermore, the study using this method confirms that forming good impressions and coming up with good ideas are based on the same essential mechanism. It is believed that this mechanism is not related to the *static nature of associations* that center around particular concepts such as "star" and "rose," but rather to the *dynamic nature of associations* that expand while a range of concepts are intricately intertwined. That is, states of "heart dancing" and "heart resonating" are not reactions to particular "seeds" but rather to the expansion and complexity of association, and the mechanism acting as a tuning fork for this is assumed to exist in the mind. If the heart dancing or resonating provides us with some kind of comfort, then we can assume that *seeking to obtain this comfortable state constitutes inner motive.*

Forming impressions is part of the post-design phase, while new concept generation is a phenomenon that occurs during processes that include the pre-design phase. This reveals the link between the pre-design and the post-design phases. Perhaps we can say that when a designer listens to their heart and is faithful to it, designing something that would make their own *heart dance*, the result is that the hearts of the users *resonate* with the designed product. This is precisely what Jobs talked about, and this is what was confirmed by these simulation experiments. In other words, if one's heart does not resonate when viewing something beautiful, one can only design dull things. We could go so far as to say that in order to move people with designs, the designer must improve his/her own sensitivity.

3.2 ONE FORM OF OUTER MOTIVE: LATENT FUNCTIONS AND LATENT FIELDS

Outer motive of design stems from products and the situations they require. How is such motive created? Based on the argument in the preceding chapter, it is born from the sharing of awareness of problems among members of the society surrounding a product and the situations that involve it. So how does this awareness of problems develop? Furthermore, can this awareness be intentionally generated? Awareness of problems arises not only from natural phenomenon like disasters, but can also be brought about by products. For instance, as products are used, new uses are uncovered, accidents occur, and thus an awareness of problems is generated leading to the design of new products. This section focuses on the awareness of problems that arises through the use of products in society. In particular, it concentrates on functions that exist latently in products and are discovered through interaction with users (these are called "latent functions" and will be explained in the next section). The section discusses the nature of such functions and considers methods for intentionally discovering them by inferring.

3.2.1 Functions and Fields

Functions of products are manifested in limited **fields**. For instance, the "driving" function of cars is only manifested when gasoline is supplied and a driver operates the vehicle on a road. Meanwhile, changing the field yields a different function. For example, when another car has broken down, a car can manifest the function of <pull car>.[1] Incidentally, a function is defined as "an entity's behavior that plays a special role." This book follows conventions for notation in the domain of design research (Pahl and Beitz, 1995) and describes functions as <verb object> or <subject verb object>.

This book focuses on the fact that products manifest different functions in different fields. General Design Theory defines the function (a) that is manifested by a product in a particular field (A) as the **visible function** of that product in field (A), and the different function (b) that is manifested by the same product in a different field (B) as the **latent function** of that product in field (A) (Yoshikawa, 1981). Expanding on these definitions, the ideas **visible field** and **latent field** have been introduced (Taura, 2014). This theory defines a field (A) in which a function (a) is manifested by a product as the visible field of that product's visible function (a), and the field (B) in which a different function (b) is manifested by the same product as the latent field of that product's visible function (a). Fig. 3.7 shows the relationship between visible functions, latent functions, visible fields, and latent fields. As an example, while the original function of a digital camera was to take commemorative pictures at scenic spots, digital cameras are now sometimes used in classrooms in place of taking notes (though the authors do not deem this desirable). Within this varied usage of a digital camera, (a) is <record scenery>, (b) is <record text>, (A) is "scenic spots," and (B) is "the classroom." Thus, the field "scenic spots" in which a digital camera manifests the function <record scenery> is the visible field for its visible

---

[1] This book will use < > to indicate specific functions. In this notation, definite/indefinite articles may be omitted to simplify these phrases grammatically.

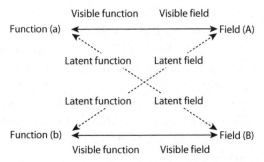

Figure 3.7 Relationship among visible function, latent function, visible field, and latent field.

function <record scenery>, and the field "the classroom" in which the digital camera manifests a different function <record text> is the latent field for its visible function <record scenery>. This means that it is possible for the photo taking function of the digital camera to not be limited to scenic spots where it was originally envisaged to be used; the digital camera can also be applicable to different fields such as the classroom.

Although it is common for latent functions and latent fields to be discovered in the process of use (interpretation) of products by users, these may also be anticipated by the designer. For instance, when a traveler realizes that they have forgotten to bring a change of socks and is forced to wash the socks they have worn and dry them with a hair dryer,[2] then such use of the hair dryer may be outside the intention of the designer. However, when a vehicle's engine stalls inside the railway boom gates (latent field) and the starter motor kicks in (the starter motor directly causes the car to move), this is within the intentions of the designer.

Furthermore, while latent functions and fields can be beneficial to users or society, they are not always so. A kitchen knife being used as a weapon is a typical example.

As mentioned, the functions manifested in each field are extremely important characteristics existing within each product.

Fields can be classified into the following three major categories: physical, contextual, and semantic.

The first are fields as physical circumstances. Products generally have intended circumstances for use, in terms of air temperature, humidity, and even voltage. Fields such as these will be called **physical fields**.

[2] This may be unsafe and requires caution.

The second are fields as scenarios of product use. An example of this is the scenario whereby a vehicle's engine stalls inside the railway boom gates and the starter motor kicks in. Such fields will be called **contextual fields**.

The third are semantic fields. For instance, in today's car-centric society (the field where cars are used), people have closer perceptions of distances than in societies where cars are rarely used. In addition, while collecting something as a hobby is meaningful to those who are interested in it, to others, it might just as well be meaningless. These kinds of fields will be called **semantic fields**.

3.2.2 Inferring Latent Functions and Latent Fields

Latent functions and latent fields are generally discovered through use of the product. If they could be inferred in advance, however, the breadth of uses of products could be expanded, and the value of use could be enhanced. This might even uncover hidden needs that lead to the development of new products. Moreover, visualizing hidden risks can help prevent accidents. In particular, reducing the hidden risks of highly dangerous products that must not be permitted to cause accidents is an important consequence, as it increases safety. Thus, if we can infer latent functions and latent fields, we can deliberately make the hidden awareness of problems related to the product and the situations surrounding it explicit. It is also expected that outer motive can be intentionally generated.

It is theorized that latent functions and latent fields can be inferred in the following ways.

First, this can be done using word replacement. Specifically, in functions described as <verb object>, the object can be replaced with something different. In the digital camera example mentioned in Section 3.2.1, the latent function <record text> is obtained by replacing the word "scenery" in the visible function <record scenery> with the word "text." The latent function of the hair dryer, <dry socks>, is also obtained by replacing the word "hair" with "socks" in the visible function <dry hair>. Another method is to maintain the object and replace the verb. If we put vegetables in a freezer, at the same time as <freeze vegetables>, this produces the function <dry vegetables>. This kind of latent function is known as a "side effect" and can be inferred by replacing the verb. Going one step further and replacing both the object and the verb makes it possible to infer a more novel latent function. The aforementioned method is achieved through searching for new words that express objects and behaviors to be inserted in the sentence. Fig. 3.8 depicts this method.

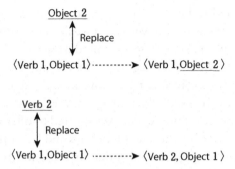

Figure 3.8 Inferring latent functions through replacement.

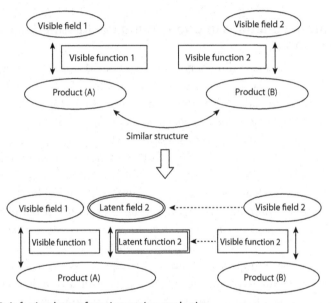

Figure 3.9 Inferring latent functions using analogies.

The second method uses the idea of analogies (Mori et al., 2013). This method is based on the idea that if a product (A) has a similar structure to product (B), then the visible function and visible field of product (B) could be the latent function and latent field of product (A) (Fig. 3.9).

For example, because a desk (A) has a similar structure to a footstool (B), we can stand on a desk when we want to reach something placed at a height. In this situation, a similar function to a footstool, <support person>, is demanded of the desk as a latent function.

Using the idea of latent field can be helpful to uncover hidden needs that lead to the development of new products. For instance, a laptop computer is normally expected to be used in a physical circumstance with a normal temperature and humidity. However, could we not use a laptop in a different circumstance? Let us imagine using a laptop during a natural disaster (latent field) such as a flood. We could imagine a computer that can work as a source of light during a blackout, receive television and radio signals, equipped with waterproofing and with a super low-energy mode for the most efficient use of battery power. Such a laptop would be expected to offer new usage value.

In this way, considering the latent functions and latent fields of products can uncover awareness of problems in the form of hidden needs for products, and this can lead to motive (in this case implicit motive) to design new products.

3.2.3 Latent Functions in Overcoming the Missing Links in Outer Motive

Until now, our understanding of products has been deepened through actual use and repeated experiences of utility and accident. In contrast, this book argues that inferring latent functions and latent fields beforehand allows the *deepening of product understanding without actual product use*. Moreover, it points to the possibility of discovering uses outside the intention of the designer *before actual use of the product*. This means that inferring latent functions and latent fields can be one powerful way of overcoming missing links in the post-design phase. Meanwhile, inferring latent functions and fields can lead to the invention (inferring) of new functions that do not result from social motive. This can *uncover awareness of problems as hidden needs, leading to the development of new products*. Thus, inferring latent functions and fields can also be a way of overcoming missing links in the pre-design phase. These two aspects of the latent functions and latent fields indicate their potential to play an important role in linking the pre- and post-design phases.

Task

Make a list of new uses (latent functions) of clothes pegs. Then, focusing on those uses, think of a new concept of product (such that the product ceases to be a clothes peg).

REFERENCES

Fellbaum, C., 1998. WordNet—An Electronic Lexical Database. MIT Press, Cambridge, MA.
Mori, H., Taura, T., Tsumaya, A., 2013. Method for inferring latent functions. In: International Conference on Engineering Design, ICED13, August 19–22, 2013.
Nagai, Y., Taura, T., Mukai, F., 2009. Concept blending and dissimilarity: factors for creative concept generation process. Des. Stud. 30, 648–675.

Pahl, G., Beitz, W., 1995. Engineering Design: A Systematic Approach. Springer, Berlin.

Taura, T., 2014. Motive of design: roles of pre- and post-design in highly advanced products. In: Chakrabarti, A., Blessing, L. (Eds.), An Anthology of Theories and Models of Design. Springer, London, pp. 83–98.

Taura, T., Yamamoto, E., Fasiha, M.Y.N., Goka, M., Mukai, F., Nagai, Y., Nakashima, H., 2012. Constructive simulation of creative concept generation process in design—a research method for difficult-to-observe design-thinking processes. J. Eng. Des. 23 (4–6), 297–321.

Taura, T., Yamamoto, E., Fasiha, M.Y.N., Nagai, Y., 2011. Virtual impression networks for capturing deep impressions, In: Gero, J. (Ed.), Design Computing and Cognition 10, Springer, pp. 559–578.

Yoshikawa, H., 1981. General design theory and a CAD system. In: Sata, T., Warman, E.A. (Eds.), Man-Machine Communication in CAD/CAM. North-Holland, Amsterdam.

PART 2

Theory and Methodology of Concept Generation

CHAPTER 4

The Metaphor Method: Theory and Methodology of Concept Generation (First Method)

Contents

4.1 THEORETICAL FRAMEWORK OF CONCEPT GENERATION USING METAPHORS

This chapter discusses the theory and methodology of **concept generation using metaphors**. When generating new concepts, we often use rhetorical techniques of comparison with existing things. If we are told to "think of a concept for any kind of new mode of transport," we find it difficult to come up with an idea. However, if we are asked to "design a mode of transport like a frog," then thinking of one becomes easier.

A number of studies have systematized rhetorical techniques. The major types of rhetorical techniques are "simile," "metaphor," "metonymy," and "synecdoche." Let us first outline each of these rhetorical techniques. The following is a very simple outline of selected content explained in references (Sato, 1992; Yamanashi, 1988).

Similes are a type of linguistic technique that describes a subject (A) by comparing it with a similar object (B). Thus, similes use phrases such as "like" and "similar to" to express the similarity, describing the subject by comparing it to another thing. This figure of speech generally appears in the form of (A is like/similar to B), and derives its name from the way in which it directly specifies such similarity using the expressions "like" or "similar to" (Yamanashi, 1988). For example, <she is like Cleopatra>.[1]

Metaphors are also a type of expression that compare their subject to a different object. Here again, there needs to be a guarantee of some kind of similarity between the subject of the metaphor (A) and the object (B). However, unlike similes, which clearly indicate similarity, using expressions such as "like" and "similar to," the similarity in metaphors is concealed behind the words. Thus, they generally take the form of (A is B) (Yamanashi, 1988). For example, <she is Cleopatra>.

Here, the difference between similes and metaphors, according to Sato (1992), is that "while similes seek to share a new understanding by explaining to the listener, metaphors expect the listener to already share an intuitive understanding." It also states that "similes are rational figures of speech, while metaphors are sensory figures of speech." However, these explanations are limited to the perspective of linguistic expressions. For the purposes of this chapter, which aims to discuss concept generation for design, there is little difference between metaphor and simile. In fact, what ever is referred to in design, as metaphor is quite often simile.

Metonyms are a type of figure of speech that express a subject using an object related to the subject. Here, the relationship in question is one of spatial adjacency or proximity, of compatibility, temporality, or causality (Yamanashi, 1988). For example, <drain one's cup>.

Synecdoches generally express a whole using a part, or a part using a whole, a species using an overarching category or vice versa (Yamanashi, 1988). For example, <ate a cow>.

Of these four techniques, metaphors can be described as a combination of synecdoche processes (Sato, 1992). For example, the metaphorical expression, <a plane is a bird>, can be interpreted as a combination of two synecdoches—one comparing the species "plane" with the category "flying thing," and the other comparing the category "flying thing" with the species "bird."

By this logic, rhetoric can be reduced to metonyms and synecdoches.

Rhetorical techniques are also often used in concept generation in design. However, it must be noted that their use in concept generation is fundamentally different from their use as linguistic expressions. In <she is Cleopatra>, "she" *already exists* when the statement is made. Whereas, in the case of design <a mode of transport like a frog>, the mode of transport to be designed *does not yet exist*. This can be more appropriately expressed

---

[1] This book will use < > to indicate specific rhetorical techniques.

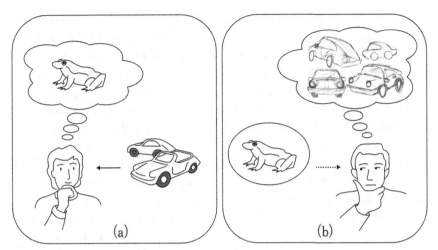

Figure 4.1 The difference between (a) expression of things that already exist using metaphors; and (b) concept generation using metaphors, where the concept does not yet exist. *(Modified from Taura and Nagai, 2012).*

as <the mode of transport I will design is a frog>. However, if an existing mode of transport were like a frog, this would be expressed as <that mode of transport is a frog>. We must note that in this way, concept generation using metaphors differs fundamentally from expressions that liken something to an existing object (Fig. 4.1) (Taura and Nagai, 2012, pp. 27–40).

In the same way, with linguistic expressions, when we use metaphors for concept generation, the key is the relationship of similarity between a frog and the mode of transport (which has no shape as it is yet to be designed). Therefore, a process is undertaken of relating a frog to modes of transport, although the two normally are not related to each other, or of generating the concept of a new mode of transport that resembles a frog. The snap-off utility knife, often cited as an example of an invention, can be explained in the same way. The idea of this type of knife is reported to have been developed with reference to blocks of chocolate (OLFA, 2014). The idea is said to have emerged in the process of trying to develop a knife that would maintain its sharp edge longer; while in this process, maintaining a sharp blade by snapping off a worn one was thought of. In this example, we can say that the idea for the snap-off utility knife was born out of the metaphor, <a new utility knife is a block of chocolate>. By focusing on a (common) property such as the ability to be snapped off, a relationship was created between a utility knife and a block of chocolate, which normally has nothing to do with each other. In other words, the concept of a new utility knife that can be snapped off like a block of chocolate (to secure a similarity between the two) was generated. In this way, concept generation using metaphors focuses on properties belonging to concepts that normally have little relationship, and generates a new concept such that this property is the same between both (to secure a similarity) (Taura and Nagai, 2012, pp. 27–40).

This kind of relativity can also be called "grouping." A (soon to be designed) new mode of transport is designed such that it is seen to be in the same group as a frog, or a (soon to be designed) new utility knife is designed such that it is seen to be in the same group as a block of chocolate. A discovery or problem is said to be groundbreaking when the grouping is adequately novel to break categories constrained within existing frameworks, and such a regrouping of categories is accepted by others (Ito, 1997). Considering this statement, we can say that the seeing of a new utility knife and a block of chocolate as the same group breaks existing categories and is also accepted by others, which makes it a groundbreaking design.

As previously discussed, metaphoric expressions signify a shared property between the subject and an existing object used in the comparison. The expression <she is Cleopatra> means that "she" and "Cleopatra" share a common property (eg, beautiful). The relationship between a concept[2] and a specific property can be expressed using set theory in terms of an element and subset relationship. For example, <she is Cleopatra> contains two elements, "Cleopatra" and "she," both of which are included in the subset "beautiful" (Fig. 4.2).

Similarly, for concept generation using metaphors, the shared property between concepts plays an important role. For example, when <a mode of transport like a frog> is to be designed, the new mode of transport will be designed with frog properties such as "hopping," and, as a result, the new mode of transport will belong to the subset where the properties of a frog are categorized (Fig. 4.3). However, this subset cannot be uniquely specified, as it is created by the designer.

These shared properties can be obtained by extracting attributes of concepts. In this chapter, we will use the term "abstraction"[3] to refer to the extraction of attributes. For example, the property of the mode of transport, "to hop," is obtained by abstracting the concept of "frog," and the property

---

[2] As discussed in Chapter 5, General Design Theory divides concepts into "entity concepts" and "abstract concepts." The concepts discussed in this chapter are entity concepts, and universal sets are entity sets as per this theory. Subsets refer to abstract concepts.

[3] Meanings of abstraction other than extraction will be discussed in Chapter 11.

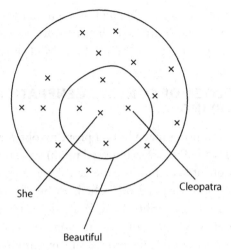

Figure 4.2 Expressions of metaphors based on set theory.

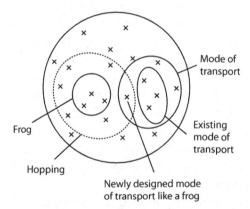

Figure 4.3 A schematic representation of concept generation using metaphors, based on set theory.[4]

of the snap-off utility knife, "to snap off," is obtained by abstracting the concept of "block of chocolate."

Metaphors can be described as a type of two-concept combining process, as described in Chapter 3. If we consider the previous examples, a

---

[4] The dotted line indicates the subset created by the designer, and therefore this boundary cannot be uniquely specified.

mode of transport like a frog is obtained by combining the two concepts of "frog" and "mode of transport," and the snap-off utility knife is obtained by combining the two concepts of "utility knife" and "block of chocolate."

4.2 METHODOLOGY OF CONCEPT GENERATION USING PROPERTY MAPPING

In concept generation using metaphors, a property obtained by abstracting a certain concept is added to (superimposed upon) an existing concept. For instance, the snap-off utility knife is generated by adding (superimposing) the property "able to be snapped off," obtained by abstracting a block of chocolate, to the traditional utility knife. A mode of transport like a frog is generated by overlaying the property "to hop," obtained by abstracting a frog, to a mode of transport. This process of superimposing properties of one concept onto another will be called "property mapping."

Design using metaphors also requires the concretization of property-mapped concepts. For instance, in the design of a mode of transport like a frog, the concept of "a mode of transport that hops," obtained from property mapping the property of a frog onto a car, must be further concretized. On the other hand, expressing existing things using metaphors only takes us until the stage of extracting properties shared by both the subject and the object of the comparison. For example, the expression <she is Cleopatra> expects only the extraction of a shared property between "she" and "Cleopatra" ("beautiful"), and nothing further.

Design using metaphors differs from linguistic expressions using rhetoric in that it is concerned with similarity with things that *do not yet exist in the real world* and that the processes of not only abstracting but also concretizing play an important role.

To summarize the aforementioned discussion, the following steps are involved in the process of design using metaphors.

Step 1: select the concepts to be used in the metaphors.
Step 2: extract (abstract) properties from the concepts selected in Step 1 (abstracting).
Step 3[5]: property map the properties extracted in Step 2, and generate a design idea by further concretizing these (property mapping and concretizing).

---

[5] Strictly speaking, the latter phrase of Step 3 is not included in the concept generation process in this book. However, this is included in this step in order to systematize this as a methodology of design using metaphors.

The method for selecting the concepts in Step 1 will be discussed in Chapter 8.

4.3 THE PRACTICE OF DESIGN USING PROPERTY MAPPING

4.3.1 Designing Artifacts Referring to Living Things

We often look toward the *natural world* for clues while working in the field of science and technology. Recently, this method has been labeled "biomimetics" or "biologically inspired design" (one example is Goel et al., 2014), and scholars worldwide have actively engaged with this method. This section introduces examples of such studies. Designing man-made objects that imitate biology means designing "man-made objects like particular living things." Thus, biomimetics is a real-world example of a method of design using property mapping.[6]

One example is the shape of the front carriage of the 500 Series Shinkansen, which was designed to imitate a kingfisher's beak. The following is a very simple outline of selected content from a website (Wild Bird Society of Japan, 2014).

The development of the 500 Series Shinkansen (bullet train), which travels at speeds exceeding 300 km/h, faced the serious problem of pressure waves that arose when the train passed through a tunnel. As the high-speed train would plunge into the tunnel, atmospheric pressure waves (compressional waves) would form, which would then grow as they moved at the speed of sound in the direction of train travel. These pressure waves would partially be released at the tunnel exit and then return in the opposite direction as expansion waves. The pressure wave thus partially released at the tunnel exit would be a low-frequency wave of 20 Hz, producing a large sound as well as far-reaching aerial vibrations. For this reason, a new shape needed to be proposed that differed from the streamlined shape that had thus far been used for the Shinkansen. During the development process, the designer wondered whether the shape of the kingfisher from bill to head could be used as a reference, given that this bird dives from low-resistance air into high-resistance water to catch its daily prey of small fish. Fig. 4.4 presents a comparison of a kingfisher and the front carriage of the 500 Series Shinkansen. There is clearly a very close resemblance. In this way, <a front carriage like a kingfisher> was designed for the 500 Series

---

[6] The example given here relates to devising a mechanism, and is not strictly concept generation but rather conceptual design. However, it is used in this chapter as an example as it provides an explanation of design using property mapping.

Figure 4.4 The front carriage of the 500 Series Shinkansen, which imitates a kingfisher.

Shinkansen. It is interesting to note that the man in charge of developing the 500 Series Shinkansen, Eiji Nakatsu, was a member of the Wild Bird Society of Japan. Incidentally, the front carriage of the 700 Series Shinkansen imitated a platypus.

Meanwhile, biomimetics-based research is gaining attention particularly in the field of material studies. Several examples of this are outlined below.

First, we provide the example of super water-repellent materials developed with reference to lotus leaves. The following is a simple explanation cited from Barthlott and Neinhuis (1997):

> In the past 25 years, scanning electron microscope studies of biological surfaces have revealed an incredible microstructural diversity of the outer surfaces of plants. Microstructures such as trichomes, cuticular folds, and wax crystalloids serve different purposes and often provide a water-repellent surface, which is not rare in terrestrial plants. Water repellency is mainly caused by epicuticular wax crystalloids which cover the cuticular surface in a regular microrelief of about 1–5 μm in height. Later, it was proven that water repellency causes an almost complete surface purification (self-cleaning effect): contaminating particles are picked up by water droplets or they adhere to the surface of the droplets and are then removed with the droplets as they roll off the leaves.

This principle has been applied to activities such as coating. This could be described as the development of <paint like a lotus leaf>.

Next, let us look at the example of adhesive material developed with reference to gecko feet. The following is a simple explanation cited from Autumn et al. (2002):

> Geckos have evolved one of the most versatile and effective adhesives known. The mechanism of dry adhesion in the millions of setae on the toes of geckos has been the focus of scientific study for over a century. A theory has been provided for dry adhesion of gecko setae by van der Waals forces, and reject the use of mechanisms relying on high surface polarity, including capillary adhesion. A van der Waals mechanism implies that the remarkable adhesive properties of gecko setae are merely a result of the size and shape of the tips, and are not strongly affected by surface chemistry. Theory predicts greater adhesive forces simply from subdividing setae to increase surface density, and suggests a possible design principle underlying the

(a)   (b)

Figure 4.5 The spaghetti tower design using metaphors (a) a tower like a tripodal ladder; and (b) a tower like a cross pole tripod.

repeated, convergent evolution of dry adhesive microstructures in gecko, anoles, skinks, and insects.

This theory has been used in the development of adhesive materials. We can consider this an example of the development of <adhesives like gecko feet>.

4.3.2 The Spaghetti Tower Design as an Example of the Use of Metaphors

To perform the spaghetti tower design task provided in Chapter 1, metaphors are often used. For instance, "tripodal ladder" and "cross pole tripod" have been used as metaphors (Fig. 4.5). All joined sections of tripodal ladders, except for the apex, are on the same level, and learning from this, joints can be made without a solid body, creating a structure that is easy to join with cello tape.

In this way, metaphors can be used to conceive outstanding structures. It should be noted that participants of this task were not instructed to design the tower using metaphors. Metaphors were used instinctively by those who undertook the task.

Figure 4.6 Five-storied pagoda (Kofukuji temple).

Incidentally, Tokyo Skytree[7] is said to have been based on the five-storied pagoda (Fig. 4.6). The five-storied pagoda is a wooden building unique to Japan. Although these pagodas have collapsed due to typhoons and fire, they are known to have superior earthquake resistance, with no record of an earthquake-induced collapse. Many theories have attempted to explain the strong earthquake resistance, but it is thought that the center column plays a crucial role. It is said that a vibration-control structure like that of the five-storied pagoda was built within Tokyo Skytree, and it was hoped that the cylindrical core (a reinforced concrete structure with a stairwell inside) built in its center would serve this function (Nikken Sekkei, 2014).

---

[7] Tokyo Skytree was built in 2012. It is a broadcasting tower with a height of 634 m. In 2011, it was recognized by the Guinness World Records as the tallest tower in the world.

4.3.3 Function Design Using Property Mapping

Requirements or specifications for new products are ordinarily provided in terms of "desired functions" (Chakrabarti and Bligh, 2001). The desired functions are generally sought in the pre-design phase, using methods such as market surveys, as outlined in Chapter 2. Within academia, while research has been conducted to describe and manipulate functions (examples are presented in Hirtz et al., 2002; Gero and Kannengiesser, 2004), little research has been conducted on generating desired functions themselves. This is probably due to the relegation of such generation of desired functions outside the realm of design, particularly engineering design. However, generating desired functions is crucial to design. This is why this book will call the generation of desired functions themselves **function design**, and approach the methodology of such design from the viewpoint of concept generation.

Since functions are types of concepts, we can use the methodology for concept generation outlined in this chapter. A simple introduction to the methodology for generation of new functions using property mapping is provided below.

As discussed in Chapter 3, functions can be described as <verb object> or <subject verb object>. Functions of a whole product will be called **whole functions**, and functions of part of a product will be called **partial functions**. For instance, the whole function of a car is <carry people>, while its partial function is <engine turns shaft>. This book will also indicate that a set of several partial functions make up the whole function of a product. In the example of a car, the whole function <carry people> is realized through a set of partial functions { <engine turns shaft>, <shaft turns tires>, …}. In order to devise a new desired function, we need not directly generate a whole function; rather, we can think of a methodology in which we first create a set of new partial functions before drawing the whole function from them (the theory and methodology of partial function manipulation in conceptual design will be discussed in detail in Chapters 9 and 10). Let us imagine replacing the partial function of a product (A) with a different product (B). This replacement means generating <product (A) like product (B)>. For example, the new desired function of a hand-operated flashlight is obtained by replacing the partial function of the battery-powered flashlight (A), <battery supplies electricity>, with the partial function of the hand-operated generator (B), <hand-operated mechanism supplies electricity> (Fig. 4.7). This manipulation is the mapping of properties from the hand-operated generator to the battery-powered flashlight at the function level.

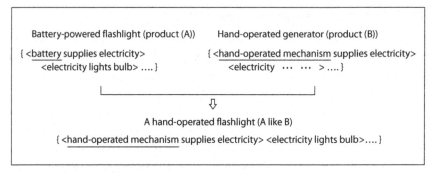

Figure 4.7 Function design using property mapping.

Based on this idea, we can systematically and innovatively design functions by formulating a methodology for property mapping of partial functions as well as the relationship between whole and partial functions (Taura and Nagai, 2012, pp. 133–149).

4.4 MULTIPLE LOOPS OF DESIGN USING METAPHORS

The keys to design using metaphors are what to use for the comparison and which property to focus on. For instance, in the example of the snap-off utility knife, what we need to ponder about is how the focus came to be upon "a block of chocolate" and "snaps off." In this section, we will consider the procedural order of appearance of these concepts. If, for example, "a block of chocolate" had come to the designer's mind first, this would imply that the property "snaps off" was derived afterward. However, "a block of chocolate" was focused upon because it could "snap off." It is difficult to believe that "a block of chocolate" came to mind first, with no relation to this property.

Did "snaps off" then come to mind first? What the designer wanted to do was to develop a knife with a long-lasting sharp edge. The idea "snaps off" is difficult to think of from the awareness of the problem "increasing the life of a blade" alone. It is more natural to think of this after seeing "a block of chocolate." Conversely, if the property "snaps off" had already been thought of, there would be no need to call to mind "a block of chocolate."

If we think of it thus, it seems unnatural for either "a block of chocolate" or "snaps off" to have been thought of first. It is more natural to assume that while thinking of a knife with a long-lasting sharp blade, the designer

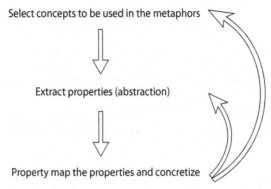

Figure 4.8 Multiple loops of concept generation using metaphors.

incidentally laid eyes upon a block of chocolate and subsequently thought about the property "snaps off." This incident shows the multiple loops of the three steps of design using metaphors, as shown in Fig. 4.8. Taking the spaghetti tower as an example, if we make it to Step 3 using "Tokyo Tower" for comparison and are not satisfied with the generated design idea, we may try a design using "a suspension bridge" for comparison. Perhaps we may *incidentally* discover an appropriate novel idea through repeating this multiple loop. Building the ability to design using metaphors requires tenacious repetition of this multiple loop.

Task

Choose five living things and five man-made objects, and combine them randomly. From this, select three pairs of "living things and man-made objects" that seem interesting. Next, think of new concepts of product using the hint <a (man-made object) like a (living thing)>.

REFERENCES

Autumn, K., et al., 2002. Evidence for van der Waals adhesion in gecko setae. Proc. Natl. Acad. Sci. 99 (19), 12252–12256.

Barthlott, W., Neinhuis, C., 1997. Purity of the sacred lotus, or escape from contamination in biological surfaces. Planta 202 (1), 1–8.

Chakrabarti, A., Bligh, T.P., 2001. A scheme for functional reasoning in conceptual design. Des. Stud. 22 (6), 493–517.

Gero, J.S., Kannengiesser, U., 2004. The situated function–behaviour–structure framework. Des. Stud. 25 (4), 373–391.

Goel, A. et al. (Ed.), 2014. Biologically Inspired Design. Springer, London.

Hirtz, J., Stone, R.B., McAdams, D.A., Szykman, S., Wood, K.L., 2002. A functional basis for engineering design: reconciling and evolving previous efforts. Res. Eng. Design 13 (2), 65–82.

Ito, M., 1997. Tacit knowledge and knowledge emergence. In: Taura, T., Koyama, T., Ito, M., Yoshikawa, H. (Eds.), The Nature of Technological Knowledge. Tokyo University Press, Tokyo (in Japanese).

Nikken Sekkei, 2014. http://www.nikken.co.jp/ensk/skytree/structure/structure_04.php

OLFA, 2014. http://www.olfa.co.jp/en/contents/cutter/birth.html

Sato, N., 1992. Retorikku Kankaku (Rhetoric Sense). Kodansya, Tokyo (in Japanese).

Taura, T., Nagai, Y., 2012. Concept Generation for Design Creativity—A Systematized Theory and Methodology. Springer, London.

Wild Bird Society of Japan, 2014. http://www.birdfan.net/fun/etc/shinkansen/

Yamanashi, M., 1988. Hiyu to Rikai (Metaphor and Understanding). University of Tokyo Press, Tokyo (in Japanese).

CHAPTER 5

The Blending Method: Theory and Methodology of Concept Generation (Second Method)

Contents

5.1 CONCEPT GENERATION USING BLENDING

Concept generation using metaphors has been discussed in Chapter 4. Concept generation using metaphors is a powerful method for generating new concepts and has been employed in a large number of actual designs. There is no doubt that this method generates new concepts, but after all, these are still "a type of utility knife," "a type of spaghetti tower," and "a type of Shinkansen." That is to say, no concept is generated by any one of these that does not fall under the categories of "utility knife," "spaghetti tower," or "Shinkansen." The newly generated concepts can be viewed as copies of "a block of chocolate," "a tripodal ladder," or "a kingfisher." If we are seeking subspecies of existing concepts, grouping-based property mapping is an effective method, but this method cannot be applied if we are seeking a concept of a completely new category. For instance, if we use the two concepts "snow" and "tomato" to create the expression "a tomato like snow," we can easily come up with the concept of "a white tomato." However, "a white tomato" is a tomato after all. Moreover, although "white," may be new in the context of tomatoes, it is an extremely ordinary property for "snow."

Here, if we were to combine the two different properties, "to fall lightly" and "acts as a flavoring," of "snow" and "tomato," respectively, we can obtain

Creative Design Engineering
http://dx.doi.org/10.1016/B978-0-12-804226-7.00005-3

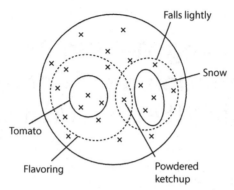

Figure 5.1 A schematic representation of concept generation using blending, based on set theory.[1]

the concept of "powdered ketchup" (which can be placed on the table while eating and sprinkled on meals as required, like powdered cheese).

This "powdered ketchup" is no longer "tomato" or "snow." Such combination of different properties of each concept to generate new concepts will be referred to as **concept generation using blending**, the theory and methodology of which will be discussed in this chapter. Fig. 5.1 shows concept generation using blending, based on set theory.

As with design using metaphors, **design using concept blending** requires further concretization of concepts. For example, design using the blending of "snow" and "tomato" can yield the idea of "powdered ketchup" by further concretizing the concept of "a tomato flavoring that falls lightly," which is obtained by combining the properties of snow and tomato.

To summarize the aforementioned discussion, the following steps are involved in the process of design using concept building.

Step 1: select the concepts to be used in the blending.
Step 2: extract (abstract) properties from the concepts selected in Step 1 (abstracting).
Step 3[2]: blend the properties extracted in Step 2, and generate a design idea by further concretizing these (blending and concretizing).

The method of selecting the concepts in Step 1 will be discussed in Chapter 8.

---

[1] The dotted line indicates the subset created by the designer, and therefore this boundary cannot be uniquely specified.

[2] Strictly speaking, the latter phrase of Step 3 is not included in the concept generation process in this book. However, this is included in this step in order to systematize this as a methodology of design using concept blending.

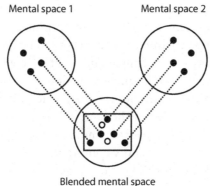

Figure 5.2 Blending of mental spaces. *(Modified from Fauconnier (1997)).*

5.2 THEORETICAL FRAMEWORK OF CONCEPT GENERATION USING BLENDING

5.2.1 Blending

In the field of cognitive linguistics, Gilles Fauconnier has conducted an analysis on mapping between mental spaces. He has shown that two input mental spaces yield a third space, which he calls a "blend." The "blend" inherits partial structure from its input spaces and has an emergent structure of its own (Fig. 5.2) (Fauconnier, 1997). Fauconnier's blending discusses mental space, but this theory can also be applied to concept generation (Nagai et al., 2009). It can be derived that two concepts generate a new concept, which inherits some properties from the first two concepts while having unique properties not found in the input concepts. This is concept generation using blending.

5.2.2 An Overview of General Design Theory

A well-known theory of concept manipulation in design is the General Design Theory of Hiroyuki Yoshikawa (Yoshikawa, 1981). General Design Theory defines design as the process of transforming a specification concept into a solution concept. General Design Theory depicts concepts as follows:

- When we experience actual existing objects by viewing or bearing some relation to them, we form concepts about these existing objects (hereafter, entities). These are **entity concepts**.
- When we possess many concepts of entities, we classify these based on some properties, creating concepts of category. An unlimited number of these can be created voluntarily by focusing on certain properties, and these are called **abstract concepts**.

The relationship between entity and abstract concepts can be expressed using set theory as an element and subset relationship, as discussed in Chapter 4. Additionally, the structure of the concept can be found using set theory or general topology. A very simple outline of selected content explained in Yoshikawa and Taura (1997) is provided below.

On seeing a dog, we remember the dog, which forms one entity concept. However, if we see some of the same animals, irrespective of whether they are dogs, sparrows, or frogs, we come to recognize the differences and commonalities between these, and we group the same animals together to form one concept of category. These category concepts are layered; for example, dogs and cats can be grouped together to create the super ordinate category concept of mammals.

Concept structure is formed in accordance with the objectives of our actions. For instance, concepts related to animals form a rigid layered structure, and such rigidity emerges from the very reason why we create the taxonomy. Dogs run fast, and so do ostriches. However, if we classify dogs as mammals and ostriches as birds, the category concept of "fast runner" that groups dogs and ostriches together is not created. Such grouping is unnecessary and is not recognized by biological taxonomy.

Although "birds" and "fast runners" are both abstract concepts from the perspective of the aforementioned categorization, they differ in the following way. "Birds" and "mammals" are concepts that have been regulated and structured in order to strictly exclude categorical ambiguities such as "a bird and a mammal." From the perspective of concept sets, this method divides concepts strictly into direct sums in mathematical terms, making it equivalent to a layered structure. As an act of categorizing, this method is correct. Meanwhile, "fast runners" is not a conventional category like the aforementioned categories of "birds" and "mammals." Furthermore, if we consider a category of animals, "loud noise makers," which consists of animals that make loud noises (although once again not used in biology), "animals that run fast and make loud noise" would clearly exist. In design, this kind of new category is often sought. However, if we introduce such categories, category concept sets will no longer be based on direct sums, and the aim of conventional categorization remains unfulfilled. Such a situation can be seen as the collapse of the layered structure

Fig. 5.3 illustrates the differences between the (a) concept structure for taxonomy and (b) concept structure for design.

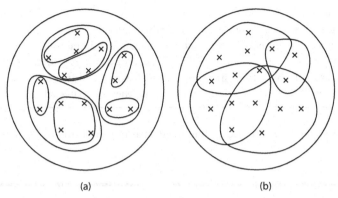

(a)   (b)

Figure 5.3 (a) Concept structure for taxonomy; and (b) concept structure for design.

Assume that designing is the act of transforming specifications[3] into solutions. In this case, specifications and solutions are both concepts. Incidentally, specifications are generally discussed with regard to functions that their designed solutions must have, and solutions are concepts in which specifications have been replaced with entities that can exist in actuality in relation to those specifications. In reality, they are sufficient information to create entities. This information should completely describe all the attributes of the entity.

Thus, concepts of functions and attributes are used in design. Function concepts are concepts that appear in the act of usage and concepts of attribute appear while analyzing a target object. These must have a more flexible composition than biological categories, as can be understood from the previous example.

This means that the design process is the replacement of specifications described by the function concept with concepts of attribute. This can be understood as a manipulation of concepts, and becomes related to the design model if we define it as mapping between sets.

To this end, the following formulation is conducted. First, several definitions are provided.

Definition 1
Entity set: a set, which includes all entities.

Definition 2
The item of attribute and its value: all entities are described completely by assigning values to a countable number of items of attribute.

---

[3] In this chapter, the term "specifications" is used in the general sense. The specific use of "specifications" in mechanical design will be explained in Chapter 9.

Definition 3

Function: when an appropriate (countable) number of items of attribute that determine an entity are selected and manifested, the total of such attributes creates one function. In this case, manifestation is something that is brought about by actions from the external world, and functions can thus be described as reactions to the actions from the external world.

Here, several axioms are provided regarding our (human) concepts.

Axiom 1

Axiom of recognition: any entity can be recognized or described by its attributes.

Axiom 2

Axiom of correspondence: entity sets and entity concept sets have one-to-one correspondence. They exhibit isomorphism.

No one possesses the concept of every entity, existing in the past, present, and future. However, this axiom is about an ideal form of knowledge. It sets such an ideal because it tries to understand real concept structures as deviations from this ideal. As a result, entity concept sets can be used in place of entity sets.

Axiom 3

Axiom of operation: the set of abstract concepts is a topology of the set of entity concepts. The topology referred to here is a phrase used in general topology in mathematics, and is defined as follows.

When set X is given a subset family ∂ with the following properties, X is called a topological space.

1. $O_\gamma \in \partial \, (\gamma \in \Gamma: \Gamma$ is a finite or infinite set$)$ is assumed to be a set family taken from within ∂. In this case, $\cup_{\gamma \in \Gamma} O_\gamma \in \partial$
2. If $O_1, O_2 \in \partial$, then $O_1 \cap O_2 \in \partial$
3. $X \in \partial$
4. ϕ (empty set) $\in \partial$

This axiom is a key to answering the question, What is a concept structure that enables design? Abstract concepts, as a category of entity concepts, are the main players in design, and what gives abstract concepts their high operability is this axiom. This is used to mathematically model the work of describing design specifications; refining them; and discovering, describing, and modifying the solution. In General Design Theory, these axioms are used to derive theorems. Some of these are listed below.

Theorem 1
Ideal knowledge is Hausdorff's space.

This is derived from the presupposition that ideal knowledge is the knowledge about everything in the real world (one-to-one). In the real world, every object can be separated from other objects. Therefore, to prevent the mixing of these concepts, any two differing entities can be separated as topologies (neighbors); in mathematics, this is a Hausdorff condition. Topological space that satisfies this condition is said to have characteristics that are very easy to separate.

Theorem 2
The specifications can be described by the intersection of appropriate abstract concepts (generally concepts of functions).

Thus, specifications become mathematically lucid. This does not necessarily need to be an intersection, and can be a union.

5.2.3 General Design Theory and Concept Generation Using Blending

The fundamental idea of General Design Theory is extremely close to that of concept generation using blending. To illustrate this, an example explained in Yoshikawa (1997) will be introduced. If we know that there are three types of meat—fresh meat, dried meat, and putrid meat—we would first recognize whether the meat is edible. From this, the abstract concepts of "things that can be eaten" and "things that cannot be eaten" arise. However, when we view putrid meat and dried meat, we also recognize that there are things that change with time and things that do not. Thus, the abstract concepts "things that change with time" and "things that do not change with time" arise. In this way, we can see how the concept abstracting process can create a number of categories for the same target set. As such, when many different categories are created, the real world can be commanded using the logical relationship between the categories. In this example, although only three types of entities exist, the world has been classified into four by introducing these two types of categories. If so, the category "things that don't change with time and can be eaten," though not existing in the real world, can be imagined in the abstract world. The category "things that don't change with time and can be eaten" did not originally exist in the real world, but these things can exist inside our minds. In other words, our minds are broader than the real world. Through the introduction of these abstract concepts, we can

[Entity]

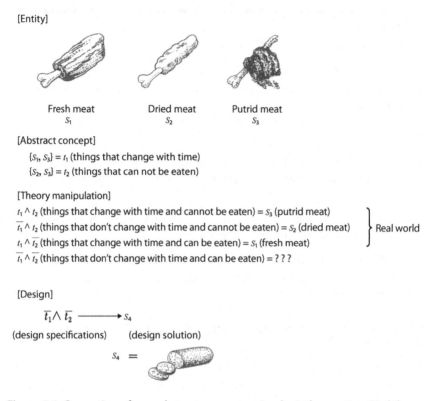

Fresh meat   Dried meat   Putrid meat
S_1 S_2 S_3

[Abstract concept]

$\{S_1, S_3\} = t_1$ (things that change with time)

$\{S_2, S_3\} = t_2$ (things that can not be eaten)

[Theory manipulation]

$t_1 \wedge t_2$ (things that change with time and cannot be eaten) = S_3 (putrid meat)

$\overline{t_1} \wedge t_2$ (things that don't change with time and cannot be eaten) = S_2 (dried meat) } Real world

$t_1 \wedge \overline{t_2}$ (things that change with time and can be eaten) = S_1 (fresh meat)

$\overline{t_1} \wedge \overline{t_2}$ (things that don't change with time and can be eaten) = ? ? ?

[Design]

$\overline{t_1} \wedge \overline{t_2} \longrightarrow S_4$

(design specifications)   (design solution)

$S_4 =$

Figure 5.4 Generation of new abstract concepts using logical operation (Yoshikawa, 1997).

generate richer concept systems in our minds than we can with real, known experiences (Fig. 5.4).

5.2.4 Concept Generation and General Design Theory

The fundamental principles of General Design Theory (Axiom 1, Axiom 2, and Axiom 3) are consistent with the perspective of this book: as concepts of new categories are generated. The example in Fig. 5.4, for instance, is in fact concept generation using blending. However, the concept generation process described in this book and the design process defined by General Design Theory seem to differ in their scope. General Design Theory defines design as a process of transforming requirements or specifications into solution in terms of concepts, and this process is situated at the beginning of the "design phase" in the framework of this book. This would mean that the concept generation process described in this book is outside the perspective of General Design Theory. However, Fig. 5.4 is an example of the pre-design

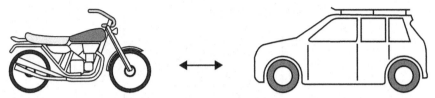

Figure 5.5 Alignable difference and nonalignable difference.

phase, confirming the principles of General Design Theory applicable to the pre-design phase. Thus, General Design Theory should be considered a theory that is widely applicable to design.

5.3 ALIGNABLE DIFFERENCE AND NONALIGNABLE DIFFERENCE

In concept generation, "usually unrelated" concepts are combined. So what does "usually unrelated" mean? This section will discuss this from the perspective of **similarities** and **dissimilarities** between concepts.

It has been found that, when engaging in concept generation, whether we focus on similarities between concepts or dissimilarities between them is a crucial factor (Nagai et al., 2009). Similarities and dissimilarities are discussed in the field of cognitive science as follows (Markman and Wisnieski, 1997; and Wilkenfeld and Ward, 2001).

Similarities and dissimilarities between concepts are categorized into three types: "commonality," "alignable difference," and "nonalignable difference." Alignable difference refers to differences along a single dimension (scale), while nonalignable difference refers to any difference between two concepts that is not along a single dimension. For example, if we compare motorbikes and cars, the difference in the number of wheels is an alignable difference, while the existence or absence of a roof is a nonalignable difference (Fig. 5.5).

Here, commonality and alignable difference are said to be more often enumerated in pairs that are seen close to rather than distant from each other. In contrast, nonalignable difference is said to be more often enumerated in pairs that are seen distant from rather than close to each other. This shows that alignable difference is a difference arising from recognition of similarity, and nonalignable difference refers to differences arising from recognition of fundamentally different mechanisms. Therefore, this book uses the term "dissimilarity" to refer to nonalignable difference.

Thereafter, the results of an experiment that indicate the relationship between recognizing dissimilarity and novelty of design outcomes (Nagai et al., 2009) will be introduced. In the design experiment outlined in

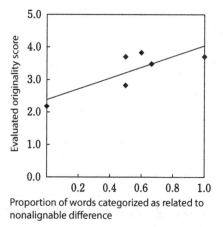

Figure 5.6 The relationship between proportion of words classified as being related to nonalignable difference and evaluated originality score of design outcomes (Nagai et al., 2009).

Section 3.1.2, the "set of words describing design outcomes" are obtained from the experiment participant. In this experiment, we have also tried to classify the "set of words explaining design outcomes" in terms of commonalities, alignable differences, or nonalignable differences between the two base concepts. For the designs that were deemed to have used blending, a positive correlation was found between the proportion of words classified as being related to nonalignable difference and the evaluated originality score[4] of the design outcome $(0.05 < p < 0.10)$ (Fig. 5.6). This result indicates that for concepts generated using blending, focusing on (nonalignable) difference between base concepts yields a design outcome with higher originality.

This result is also consistent with General Design Theory. Theorem 1 identifies concept space as ideal, what mathematics calls Hausdorff space (a type of separated space). This means that, for design, concept space is formed in the manner in which dissimilarity is recognized.

If we refer to the example of "snow" and "tomato" in Section 5.1, the "lightly falling property" of "snow" and the "used as a flavoring property" of "tomato" are related by nonalignable difference. Whereas, in "tomato like snow," when the "white" property of "snow" is mapped onto "tomatoes," the "red" of "tomatoes" is replaced with "white"—"white" and "red" are related by alignable difference.

---

[4] Eleven raters evaluated the design outcomes on a five-point scale (with 5 being the highest), and the average of these was taken as the evaluated originality score. For details about the experiment, see Nagai et al. (2009).

Figure 5.7 Example of a design outcome using concept blending.

From this discussion, we can gather that concept generation using metaphors makes more use of property mapping based on alignable difference, while concept generation using blending makes more use of blending based on nonalignable difference.

5.4 THE PRACTICE OF CONCEPT GENERATION USING BLENDING

5.4.1 An Example of Design Using Concept Blending

The design outcome from the design experiment (Nagai et al., 2009), thought to have been designed using concept blending, is shown in Fig. 5.7. This example was designed by blending the concepts of "ship" and "guitar," and is explained thus, "sound is made by using the propulsion of the ship and waves on the water surface to vibrate strings. This ship can be hired out at leisure centers, where customers can ride it around as an instrument and can even play duets." This idea is a combination of the properties of "ship" and "guitar," and involves coming up with something that goes beyond both of these categories.

An actual example of a design that can be interpreted as having used concept blending is the "space shuttle." Although it may seem like an afterthought, the space shuttle can be considered a combination of the properties of both an "airplane" and a "rocket." A "mobile phone with inbuilt

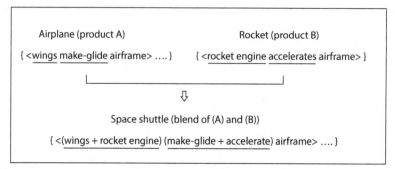

Figure 5.8 Function design using blending.

camera" can also be interpreted as having been designed by blending the concepts of "telephone" and "camera."

5.4.2 Function Design Using Blending

This section will discuss the methodology for designing new desired functions using blending.

As in Chapters 3 and 4, functions will be described using sets of verbs and nouns, that is as <verb object> or <subject verb object>. In addition, the functions of a whole product are referred to as "whole functions," and functions of part of a product as "partial functions," as defined in Section 4.3.3. Let us discuss the methodology of creating a new set of partial functions that can generate new whole functions. Assume, for example, that some of the partial functions of a product (A) are similar to some of the partial functions of a different product (B). In this case, let us form a union by sharing the partial functions that are the same or similar between both products, and think of a new set of partial functions. Then, a new whole function can be derived from this newly created set of partial functions. In this way, we can design new desired functions for product (A) and product (B) using blending. For example, in the space shuttle example, by blending the partial function of an airplane (product A), <wings make-glide airframe>, with the partial function of a rocket (product B), <rocket engine accelerates airframe (through space)>, through the common "airframe," we can describe the partial function <(wings and rocket engine) (make-glide and accelerate) airframe> (Fig. 5.8). While determining similarities among some partial functions and sharing each between different products, if we can determine not only whether the words used to describe functions match syntactically but also whether their meaning is the same as well, then we will be able to blend functions more flexibly. Additionally, by formulating the relationships between whole and partial functions and a methodology for blending partial functions, we can design functions systematically (Taura and Nagai, 2012).

Task

Design a new product blending the two concepts of firefly and rice.

REFERENCES

Fauconnier, G., 1997. Mappings in Thought and Language. Cambridge University Press, Cambridge, United Kingdom.

Markman, A.B., Wisnieski, E.J., 1997. Similar and different: the differentiation of basic-level categories. J. Exp. Psychol. Learn. Mem. Cognit. 23, 54–70.

Nagai, Y., Taura, T., Mukai, F., 2009. Concept blending and dissimilarity: factors for creative concept generation process. Des. Stud. 30, 648–675.

Taura, T., Nagai, Y., 2012. Concept Generation for Design Creativity: A Systematized Theory and Methodology. Springer, Berlin, Germany, pp. 133–149.

Wilkenfeld, M.J., Ward, T.B., 2001. Similarity and emergence in conceptual combination. J. Mem. Lang. 45, 21–38.

Yoshikawa, H., 1981. General design theory and a CAD system. In: Sata, T., Warman, E.A. (Eds.), Man-Machine Communication in CAD/CAM. North-Holland, Amsterdam.

Yoshikawa, H., 1997. A new paradigm of engineering. In: Taura, T., Kayama, T., Ito, M., Yoshikawa, H. (Eds.), Topologies of Technological Knowledge. University of Tokyo Press, Tokyo, (in Japanese).

Yoshikawa, H., Taura, T., 1997. Process knowledge of general design theory. In: Taura, T., Kayama, T., Ito, M., Yoshikawa, H. (Eds.), Topologies of Technological Knowledge. University of Tokyo Press, Tokyo, (in Japanese).

CHAPTER 6

Using Thematic Relations: Theory and Methodology of Concept Generation (Third Method)

Contents

6.1 TAXONOMIC AND THEMATIC RELATIONS

Generally, two types of relationships are said to exist between concepts: "taxonomic relation" and "thematic relation" (Wisniewski and Bassok, 1999). The former type of relationship focuses on the similarity of properties between the two concepts, while the latter indicates the relationship between the concepts through a *thematic scene*. For example, as shown in Fig. 6.1, "an apple" and "a mandarin" have shared properties such as "round" and "edible"; therefore, the two have a taxonomic relation. On the other hand, "an apple" and "a knife" have different shapes and are made of different materials. There are almost no shared properties between the two. However, it is not uncommon to see an apple and a knife placed together. This is because we know that <apple is cut by knife>.[1] In this way, even when two concepts have little to no taxonomic similarity, they can be perceived as having a close relationship. Such a relationship is referred to as a thematic relation.

---

[1] This book will use < > to indicate specific thematic relations. In this notation, thematic relations will be described as <noun 1, verb that links noun 1 and noun 2, noun 2> and definite/indefinite articles may be omitted to simplify these phrases grammatically.

Creative Design Engineering
http://dx.doi.org/10.1016/B978-0-12-804226-7.00006-5

71

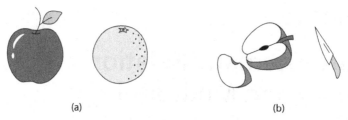

(a)   (b)

Figure 6.1 (a) Taxonomic relation; and (b) thematic relation.

What kinds of specific relationships exist within a thematic relation? The following relationships illustrate thematic relations (Shoben and Gagne, 1997; Georgiev et al., 2016):

- A relationship between an object and a place
  This is the thematic relationship between an object and the place where the object exists, for example, <music is listened to on train>.
- A relationship between the whole object and one of its parts
  This is the relationship between a whole object and one of the parts that it is composed of, for example, <engine accelerates car>.
- A relationship between a product and its material
  This is the relationship between a product and one of the materials from which it is made, for example, <bread is made from flour>.
- A relationship between causes and effects
  This is the relationship between the cause and effect of a phenomenon, for example, <engine turns tires>.
- A relationship between a tool or method, and its target object
  This is the relationship between a tool or method and the object it affects, for example, <driver turns screw>.

If we view these relationships from the perspective of the rhetoric technique outlined in Chapter 4, the relationship between the whole and the partial corresponds to synecdoche, and the remainder of relationships correspond to metonymy. On the other hand, taxonomic relations are equivalent to synecdoches. Thus, the difference between a taxonomic relation and a thematic one largely corresponds to the difference between synecdoche and metonymy.

6.2 THE ROLE OF THEMATIC RELATIONS IN DESIGN

What role does a thematic relation play in design? The following lists some possibilities.

- Situations in which products are used are sometimes described using thematic relations. For instance, a situation in which a portable music player is used is described as a thematic relation, for example, <music is listened to on train>.

- Product interfaces are sometimes described using thematic relations. For instance, methods for operating multifunction portable terminal screens are described using a thematic relation, for example, <screen is scrolled using one's finger>.
- Functions of products are described using thematic relations. For instance, the linguistic form <subject verb object>, for example, <engine turns shaft>, is a thematic relation in itself.
- The internal structure of products (the relationship between the whole and its parts) is described using thematic relations. For instance, the structure of a car is described in terms of a thematic relation, for example, <car has tires>.

These relationships cover almost all phases of design. Let us focus in particular on the description of situations for product use and features of the products themselves (structure and function) using thematic relations. These relationships correspond to the field and function relationships mentioned in Chapter 3. In fact, situations of use of a product are associated with fields. This chapter will primarily discuss methods for creating new situations using thematic relations.

6.3 METHODOLOGY AND PRACTICE OF DESIGN USING THEMATIC RELATIONS

6.3.1 Methodology of Design Using Thematic Relations

Focusing on the relationship between situations of the use of a product and features of the products themselves and describing these using thematic relations allows us to think of methods for generating new concepts of product (this will be called **concept generation using thematic relations**). This method first involves the (before determining the features of the product to be used) creation of a new situation using a thematic relation, and then referring to such a situation to generate a new concept of product (which is useful in the situation, or which can be realized as a function of that situation). Let us consider the "snow" and "tomato" example. First, let us say that from "snow" and "tomato," <tomatoes are preserved in snow> comes to mind. We may notice that high moisture in snow will preserve the freshness of the tomatoes. Moreover, if we recall that one limitation of today's refrigerator is that it dries out food, we can devise a new concept (function) for a refrigerator, <keep food with moisture>. This is concept generation using a thematic relation. In this example, the new situation

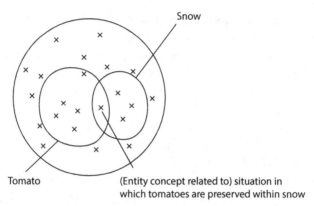

Figure 6.2 A schematic representation of concept integration in thematic relation, based on set theory.

was thought of ahead of and independently from the concept of product (function), and both of them (situation and function) were described using a thematic relation.

Here, let us consider the earlier stage in the process of generating new concepts using thematic relations in more detail. As the foregoing example involves integrating two concepts, what is happening is a two-concept combining process called **concept integration in thematic relation** (Nagai et al., 2009). In set theory, if we interpret the intersection of subsets as describing a thematic relation of those subsets (abstract concepts), then we can describe the process of concept integration in thematic relation as an operation on subsets (Fig. 6.2).

In this way, new situations for the use of a product can be created through concept integration in thematic relation. In addition, new situations can be also inferred as follows. One method is to replace one word in the known situation that is described using a thematic relation with a different word. For example, the situation <music is listened to within (on) train> can be obtained from the known situation <music is listened to within room> by replacing the word "room" with "train." Another method is integrating multiple known situations. For example, to obtain the situation <scenic photo is sent using mobile phone> (the so called "picture message"), first the two situations <message is sent using mobile phone> and <photographic postcard is sent by mail> are integrated to create the situation <photographic postcard is sent using mobile phone>, and then the word "photographic postcard" is replaced by a word with the same meaning, "scenic photo" (Fig. 6.3).

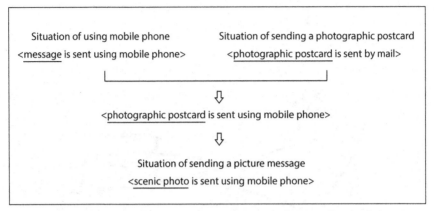

Figure 6.3 Generation of a new situation through combining of multiple situations described using thematic relations.

Generally, the manner in which situations described using thematic relations are integrated does not matter. We will, however, now focus on those that describe aspects of our daily lives. Even new situations are often modeled on something from daily life. We can therefore obtain new situations by reconstructing situations that exist within our daily lives. In the aforementioned example, the situation <photographic postcard is sent by mail> has been described in daily life, and the situation <message is sent using mobile phone> is a description of the latest technological method for reconstructing this. In order to actualize this method, we need to prepare a database that stores a large number of examples that are described using this thematic relations (Georgiev et al., 2016).

Next, let us consider the later stage in the process of generating new concepts using thematic relations. There are two types of methods for doing this. The first involves devising a concept of product that can actualize a newly created situation as a function, such as the situation <tomato is preserved in snow>, in the example of refrigerators that can <keep food with moisture>. The second method involves devising concepts of product that are useful in newly created situations, such as <music is listened to within (on) train>, in the example of music players that are small enough to be carried within (on) train. Both methods, including the process of concretization of the generated concept, are referred to as **design using thematic relations**.

To summarize the aforementioned discussion, the following steps are involved in the process of design using thematic relations.

Step 1: select the concepts to be used in the integration.

Figure 6.4 Example of a design outcome using a thematic relation.

Step 2: integrate the selected concepts thematically and create a new situation.

Alternatively, a new situation can be created by using the replacing or integrating methods instead of Steps 1 and 2. Additionally, applying the repeated replacement or integration for what has already been created using known situations may enable the generation of even more novel situations.

Step 3[2]: generate a new concept of product that is useful in the newly created situation or devise a concept of product that is realized as a function of that situation, and concretize these.

The method for selecting the concepts in Step 1 will be discussed in Chapter 8.

6.3.2 The Practice of Design Using Thematic Relations

The design outcome from the design experiment (Nagai et al., 2009) thought to have been designed using thematic relations is shown in Fig. 6.4. This example was designed by integrating the concepts "ship" and "guitar," and explained as, "being on a ship puts you in a good mood and you want to sing. When this happens, it would be great to have a guitar that is easy to

[2] Strictly speaking, the latter phrase of Step 3 is not included in concept generation process in this book. However, this is included in this step in order to systematize this as a methodology of design using thematic relations.

play even when the ship sways a lot. This guitar sticks to your body in more places and fits better than a regular guitar. The guitar can be fixed to your torso such that even when you lose your balance, you can still use both your hands quickly." We can interpret this example as the designer having first integrated the two concepts "ship" and "guitar" using the thematic relation <guitar is played on ship>, before devising the concept of an instrument that can be used (played) in the situation described in the thematic relation (a situation in which one is swaying).

An example of an actual design using thematic relations is "a portable music player." Until this invention, music was something to be listened to indoors. We can interpret this design process as someone envisaging a situation in which one could "listen to music anywhere (eg, on a train)," which was then followed by the devising of a concept of product (a portable music player) that enabled its use in this situation.

6.4 FUNCTION DESIGN USING THEMATIC RELATIONS

This section discusses the methodology for designing new desired functions using thematic relations.

In addition to the methods given in Section 6.3.1, that is, thinking of newly created situations as new functions or devising functions that are useful in the newly created situations, function design using thematic relations can also be achieved using the ideas of latent functions and latent fields. This is because, as discussed in Chapter 3, situations in which products are used are associated with fields, and latent functions are nothing but new functions.

Let us consider the "clothes pegs" task provided in Chapter 3. Let us assume that the latent function of "clothes pegs" we came up with was <stop the bleeding (from a finger)> of a "rubber band" and <clip-together paper> of a "clip." These fields (latent fields) are <finger is injured with bleeding> and <documents are organized in a file>. From this, we can think of new product concepts (functions) for clothes pegs such as "rubber bands" and "clips" (which need no longer be clothes pegs). Furthermore, we can also apply the method discussed in Section 6.3.1 to these latent fields. For example, if we create the situations <documents are classified into some categories (without opening them)> from <documents are organized in a file>, we can think of products with the functions <write easily index> and <link-together paper>, such as labels and clips like magnets.

In order to support the design of new functions using thematic relations, we must prepare a database of relationships between functions and fields. If we use the example of the "clothes pegs," the database would store information as follows: the (latent) function <stop bleeding (from a finger)> of a "rubber band" is useful in the situation <finger is injured with bleeding>, and the function of <write index> is useful in the situation <documents are classified into some categories>.

Additionally, describing fields (situations) and functions in a unified manner makes searching for examples and manipulating these easier. For example, the field (situation) "cut an apple with a knife" can be described in three items as <apple is cut by knife>. Functions can also be described in three items as <subject, verb, object>. If we describe both using the same format, we can systematize these relationships to effectively infer new product functions from situations in which products are used.

Task 1

Think of a situation that includes the three keywords "skyscraper," "banana," and "pencil." Next, think of a new concept of product based on this situation.

Task 2

Think of a new function integrating "car," "umbrella," and "spring." You may use other products also. You may also separate the products into parts to use them, and you do not need to use all the parts.

Task 3

Choose a situation from daily life and describe it using a thematic relation. Combine this with a thematic relation that describes an advanced technological method, and generate a new situation. Next, think of a new concept of product based on the newly created situation.

REFERENCES

Nagai, Y., Taura, T., Mukai, F., 2009. Concept blending and dissimilarity: factors for creative concept generation process. Des. Stud. 30, 648–675.

Shoben, E.J., Gagne, C.L., 1997. Thematic relation and the creation of combined concepts. In: Ward, T.B., Smith, S.M., Vaid, J. (Eds.), Creative Thought: An investigation of Conceptual Structures and Processes. APA Book, Washington.

Georgiev, G.V., Sumitani, N., Taura, T., 2016. Methodology for creating new scenes through the use of thematic relations for innovative designs. Int. J. of Design Creativity and Innovation (published online).

Wisniewski, E.J., Bassok, M., 1999. What makes a man similar to a tie? Stimulus compatibility with comparison and integration. Cognit. Psychol. 1, 208–238.

CHAPTER 7

Abduction and Concept Generation

Contents

7.1 DEDUCTION, INDUCTION, AND ABDUCTION

In this chapter, concept generation will be examined from the perspective of *reasoning*. Generally speaking, there are three types of reasoning: "deduction," "induction," and "abduction." The characteristics of each will be explained below.

Deduction is the reasoning that derives specific knowledge from general or universal knowledge, and syllogism is the typical example of this. When first-order predicate calculus is used in describing the reasoning process, deduction is formulated as in Fig. 7.1. Here, $\forall$ is a type of quantifier called the universal quantifier, and means "for all individuals—." There is another quantifier that will not be used in this chapter, $\exists$, which means "there is at least 1 individual that is—" (the existential quantifier). In addition, X is called the variable, and represents an unspecified individual. "Human" and "mortal" are predicates, and "Socrates" is an object constant. $\rightarrow$ is a type of logical connective, and is read as "implies" (in layman's terms, "if—, then—"). Other logical connectives include $\land$ (conjunction: in layman's terms, "and"), and $\lor$ (disjunction: in layman's terms, "or").

Fig. 7.1 shows the flow of reasoning from the general or universal knowledge <all humans are mortal>[1] and the specified knowledge <Socrates is (was) human>, to the knowledge that therefore <Socrates is (was) mortal>.

---

[1] This book will use < > to indicate knowledge expressed in the format of reasoning. Definite/indefinite articles may be omitted to simplify these phrases grammatically.

Creative Design Engineering
http://dx.doi.org/10.1016/B978-0-12-804226-7.00007-7
79

$$\forall X \, (\text{human}(X) \rightarrow \text{mortal}(X))$$
$$\text{human}(\text{Socrates})$$
_____
$$\text{mortal}(\text{Socrates})$$

Figure 7.1 Formal description of deduction.

$$\text{human}(\text{Socrates})$$
$$\text{mortal}\,(\text{Socrates})$$
_____
$$\forall X \, (\text{human}(X) \rightarrow \text{mortal}(X))$$

Figure 7.2 Formal description of induction.

Induction is the reasoning that derives the general or universal knowledge underpinning the given (observed) experiential or specified knowledge. Fig. 7.2 illustrates induction using first-order predicate calculus.

In this example, if the specified knowledge <Socrates is (was) human> is provided beforehand, and the experiential knowledge <Socrates is (was) mortal> is provided as well, then the general or universal knowledge <all humans are mortal> is induced.

Abduction is the reasoning whereby we discover hypotheses that explain certain facts that we encounter. Fig. 7.3 illustrates abduction using first-order predicate calculus.

This example shows how the specified knowledge <Socrates is (was) human> is abducted from the experiential knowledge <Socrates is (was) mortal> and the general or universal knowledge <all humans are mortal>.

$$\text{mortal}(\text{Socrates})$$
$$\forall X(\text{human}(X) \rightarrow \text{mortal}(X))$$
_____
$$\text{human}(\text{Socrates})$$

Figure 7.3 Formal description of abduction.

7.2 ABDUCTION AND DESIGN

7.2.1 Hypothesis Generation and Abduction

Which type of reasoning does design employ—deduction, induction, or abduction? To answer this question, this section will introduce Yoshikawa's discourse (Yoshikawa, 1997). This discourse states that the process of deriving laws and the act of applying theories (laws) are in a sense the same.

Indeed, it states that the generation of hypotheses and the act of designing has the same logical structure. Of course, this merely means that their logical structure is the same, and not that they are the same act. The generation of hypotheses, that is, the generation of laws, begins with observation. Observation is conducted based on a fixed viewpoint, and collections of observations about phenomenon are made. If we assume that it is possible to count the entire phenomenon that we are trying to include within our hypotheses, then we can create a group of descriptions of phenomenon A. For example, assume that a law B exists such that if B is true, then A is definitely true. Deriving B thus is called abduction.

On the other hand, the act of applying the law can also be considered abduction. Let us consider the example of house design, in which a designer thinks of a house based on a client's requirements. In this case, describing the requirements of a client can be considered to correspond to describing the phenomenon observed in a particular field (A), and the house can be considered to correspond to the laws (B). Therefore, the structure is such that if B is true, then A is true (Fig. 7.4).

Based on this consideration, designing a house that satisfies the client's requirements or specifications is abduction in the sense that it has the same logical structure as the generation of laws (hypotheses) that explains observed phenomenon.

Hypothesis generation
  A: Phenomenon observed in a particular field
  B: Laws that can explain every phenomenon in that field

        Derivation of B: if A is given, then B →A

Design
  A: A client's requirements for designing a house
  B: A house thus designed that it satisfies all of the client's requirements

        Derivation of B: if A is given, then B →A

Figure 7.4 Similarity between the logical structure of hypothesis generation and design (Yoshikawa, 1997).

7.2.2 Concept Generation and Abduction

In reasoning, even when the knowledge that forms an assumption is *common knowledge*, the knowledge derived from this can be common or *odd knowledge*. This section will discuss this point. Predicate calculus deals with truth or falsehood; this section will however focus on whether the content of the knowledge is "common" or "odd." Here, "common knowledge" means that the knowledge is uniquely interpreted and that such an interpretation is judged to be universally true. This differs in the case of "odd knowledge." We will focus on this perspective in order to discuss abduction in relation to the metaphors mentioned in Chapter 4. Metaphors are sometimes odd expressions that are not common knowledge. For example, the expression <a new car is a frog>, strictly speaking, is false. This is because a "car" is not a "frog." However, this metaphor has the possibility of generating novel knowledge. The kinds of knowledge that will be generated from each of the three types of reasoning will be discussed below.

First, let us consider deduction. In Fig. 7.1, we can replace "human," "mortal," and "Socrates" with any kind of predicate or object constant. In the case that the knowledge in the assumptions (the two formulae described above the horizontal line in the figures) is common knowledge, let us judge whether the knowledge derived from these formulae is common or odd knowledge. If we actually attempt this, as shown in Fig.7.5, we find that in deduction, common knowledge always generates common knowledge.

Now, let us consider induction. Similarly, in Fig. 7.2, we can replace "human," "mortal," and "Socrates" with any kind of predicate or object constant. In the case that the knowledge in the assumptions (the two formulae described above the horizontal line in the figures) is common knowledge, let us judge whether the general knowledge derived from these formulae is common or odd knowledge. If we actually attempt this, as shown in Fig. 7.6, we find that in induction, both common knowledge (such as <all fruits are edible>) and odd knowledge (such as <all humans are geniuses>) can be generated.

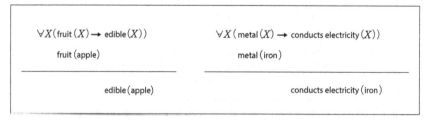

Figure 7.5 An example of knowledge generated by deduction.

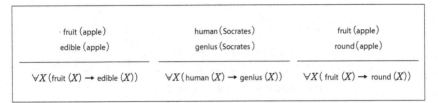

Figure 7.6 An example of knowledge generated by induction.

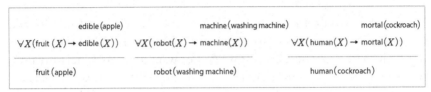

Figure 7.7 An example of knowledge generated by abduction.

Finally, let us consider abduction. Similar to deduction and induction, in Fig. 7.3, we can replace "human," "mortal," and "Socrates" with any kind of predicate or object constant. In the case that the knowledge in the assumptions (the two formulae described above the horizontal line in the figures) is common knowledge, let us judge whether the knowledge generated is common or odd knowledge. If we actually attempt this, as shown in Fig. 7.7, we find that in abduction, both common knowledge (such as <apples is fruit>) and odd knowledge (such as <washing machine is robot> or <cockroach is human>) can be generated.

However, what kind of knowledge is referred to by odd knowledge? Strictly speaking, odd knowledge is false knowledge. However, it may include novel ideas. For example, the expressions in Fig. 7.7, <cockroach is human> and <washing machine is robot>, strictly speaking, are false. Cockroaches are not humans, nor are washing machines robots. However, if we view these as metaphorical expressions of knowledge, this can be (broadly) interpreted as the implicit expression of the kinds of properties of cockroaches and washing machines.

How is design described using first-order predicate calculus? For example, the conception of the structure of the spaghetti tower presented in Chapter 4 by taking a hint from a "tripodal ladder" can be described as abduction, as shown in Fig. 7.8. Here, it must be noted that the expression <the new spaghetti tower is a tripodal ladder>, although odd, is valid as a metaphor. In actuality, mimicking the shape of a "tripodal ladder" is not false from an engineering perspective. This is because it makes the spaghetti easier to join.

$$\frac{\text{sturdy structure (new spaghetti tower)}}{\forall X \, (\text{tripodal ladder}(X) \to \text{sturdy structure } (X))}$$
$$\text{tripodal ladder(new spaghetti tower)}$$

Figure 7.8 Description of the design process using abduction.

The aforementioned discussion shows how the pursuit of novelty and the making of mistakes are two sides of the same coin. Odd knowledge is not derived by deduction. Thus, if we do not want to make mistakes, deduction should be used. However, if we do so, no novel ideas will be generated either. On the other hand, if we want to create novel ideas, we must engage in induction or abduction; however, this means that we risk making mistakes. In other words, this discussion reiterates that "nothing novel can be achieved if we fear failure," or conversely, "if we do something new, it is possible we will fail."

7.2.3 Content and Abduction

Let us exchange "sturdy structure" and "tripodal ladder" in the general and universal knowledge in Fig. 7.8. This transformed knowledge is described in the rearranged expression in Fig. 7.9.

This refers to deduction. The comparison between Figs. 7.8 and 7.9 illustrates how the type of reasoning (deduction or abduction) is determined. It can be observed that the direction of the arrow (implication) relating the two propositions that comprise the general or universal knowledge determines the type of reasoning. The general or universal knowledge in Fig. 7.8 illustrates that, in layman's terms, if it is a "tripodal ladder," then it is a "sturdy structure," meaning that there are "sturdy structures" other

$$\frac{\forall X (\text{sturdy structure } (X) \to \text{tripodal ladder } (X))}{\text{sturdy structure (new spaghetti tower)}}$$
$$\text{tripodal ladder (new spaghetti tower)}$$

Figure 7.9 Transformation of general or universal knowledge in abduction.

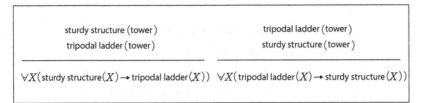

Figure 7.10 Direction of implication of knowledge generated using induction.

than "tripodal ladders." Meanwhile, what the knowledge corresponding to this is in Fig. 7.9 indicates that, if it is a "sturdy structure," then it is a "tripodal ladder," and thus there are no "sturdy structures" that are not "tripodal ladders."

So, what determines the direction of implication for each of the two propositions? General or universal knowledge is generated through induction. However, in the case of induction, the direction of implication for the general or universal knowledge is not determined by the knowledge in the assumptions (the two formulae described above the horizontal line in the figures). As shown in Fig. 7.10, the formula allows for knowledge with implication in either direction to be generated. This illustrates that, in the formula, the direction of the implication that links the two propositions cannot be determined by objective knowledge such as the laws of physics; instead, it is determined by the content of the propositions themselves. Therefore, whether an act of reasoning is abduction or deduction is not determined formally (superficially), but by the content of the knowledge in the assumptions.

As outlined just earlier, in the case of deductive reasoning, common knowledge always generates common knowledge. However, for example, if the (personal) common sense of the designer did not match that of the general society, then there is the possibility that deduction is mistakenly used instead of abduction. If this occurs, there is a possibility that an inaccuracy that deserves careful consideration might be mistakenly considered true (and can no longer be perceived as a mistake). For example, as shown in Fig. 7.11, if we apply this to the scenario of designing a sturdy desk, the inadequacy of the desk goes unnoticed. In other words, even for an explanation that at first glance seems to hold syllogistic weight, if the general or universal knowledge that forms its assumptions is not definitely common knowledge, then the logic itself becomes meaningless, and we must be careful not to make inappropriate assertions.

$$\frac{\begin{array}{c} \forall X \left(\text{sturdy structure}\left(X\right) \; \rightarrow \; \text{tripodal ladder}\left(X\right) \right) \\ \text{sturdy structure}\left(\text{new desk}\right) \end{array}}{\text{tripodal ladder}\left(\text{new desk}\right)}$$

Figure 7.11 An example of reasoning using deduction where abduction should be used.

7.3 THE RELATIONSHIP BETWEEN ABDUCTION AND METAPHOR

The previous section outlined how abduction can be interpreted as metaphor. That is, it showed how knowledge derived from abduction can be interpreted as a metaphor. To add to this, this section will now explain how the structure of reasoning using abduction is similar to the structure of concept generation using metaphors (Taura and Nagai, 2013).

If we apply the formula of abduction to the example of the utility knife given in Chapter 4, it can be described as shown in Fig. 7.12. Here, if we interpret the predicate as an abstract concept (property) of an object, then formula (1) illustrates how the (entity) concept "new utility knife" belongs to the abstract concept "snap off," and formula (2) shows how the abstract concept "block of chocolate" is included in the abstract concept "snap off." Finally, formula (3) shows how the (entity) concept "new utility knife" belongs to the abstract concept "block of chocolate."

Let us express the substances of Fig. 7.12 using set theory. However, because formula (3), strictly speaking, is false, "new utility knife" will be pictured as not being included in "block of chocolate." If we do so, we obtain a diagram like Fig. 7.13. This diagram is roughly the same as Fig. 4.2, which showed the structure of metaphors. This illustrates the fact that the structure of abduction is extremely close to that of metaphor.

$$\frac{\begin{array}{ll} \text{snap off}\left(\text{new utility knife}\right) & (1) \\ \forall X \left(\text{block of chocolate}\left(X\right) \; \rightarrow \; \text{snap off}\left(X\right) \right) & (2) \end{array}}{\begin{array}{ll} \text{block of chocolate}\left(\text{new utility knife}\right) & (3) \end{array}}$$

Figure 7.12 Description of the process of generating the idea for the snap-off utility knife using abduction.

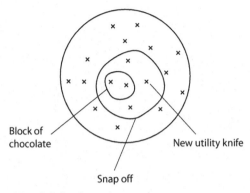

Figure 7.13 Expression of abduction using set theory.

Accordingly, we can see how metaphor and abduction are similar in that they rely on similar (common) properties shared between two ordinarily unrelated concepts, and generate new concepts in order to secure the shared similar properties. They differ in that the shared properties between the concepts are not identified in the case of metaphors, while they are explicitly identified within the knowledge in the assumptions in the case of abduction.

Task

Create at least two examples each of deduction, induction, and abduction. For the examples of induction and abduction, verify that both common and odd knowledge can be derived, even if the assumptions are formed of common knowledge.

REFERENCES

Taura, T., Nagai, Y., 2013. A systematized theory of creative concept generation in design: first-order and high-order concept generation. Res. Eng. Des. 24 (2), 185–199.

Yoshikawa, H., 1997. A new engineering paradigm as historical science. In: Taura, T., Koyama, T., Ito, M., Yoshikawa, H. (Eds.), Topologies of Technological Knowledge. University of Tokyo Press, Tokyo (in Japanese).

CHAPTER 8

Basic Principles of Concept Generation

Contents

8.1 THE VALIDITY OF CATEGORIZING CONCEPT GENERATION METHODS INTO THREE TYPES

The three methods of concept generation outlined in Chapters 4–6 are all two-concept combining processes. For instance, in the design <a tomato like snow> example, the properties of "snow" are mapped onto "tomato." This property mapping process can be viewed as a process of combining the two concepts "snow" and "tomato." Similarly, the two concepts, "snow" and "tomato," are combined when using the blending method or the thematic relations method. In all the methods, the two concepts being combined will be referred to as "base concepts," in the same way as in Chapter 3.

Meanwhile, in the field of linguistics, research is being conducted on the *interpretation* of noun–noun phrase pairs. According to this research, noun–noun phrase pairs are interpreted based on three methods, "property mapping," "hybrid," and "relation linking" (Wisniewski, 1996). For example, a knife–fork is interpreted by property mapping as a knife-shaped fork; by hybrid as half knife, half fork; or by relation linking as a knife and fork set from the context of their use at meal times (Fig. 8.1). Here, let us consider the two nouns that make up the noun–noun phrase pair as the two base concepts for use in the two-concept combining process. If we do so, as shown in Table 8.1, we can see that each of these corresponds to one of the three concept generation methods. This shows that the types of concept generation in design are consistent with the types of nouns–noun phrase pairs in linguistics, providing a basis for the categorization of concept generation methods into the metaphor method, the blending method, and the thematic relations method.

Creative Design Engineering
http://dx.doi.org/10.1016/B978-0-12-804226-7.00008-9

Knife-shaped fork

Half fork, half knife

Knife and fork set

Figure 8.1 Three types of interpretation of a noun–noun phrase pair.

Table 8.1 Comparison of the two-concept combining process and interpretation of noun–noun phrase pairs

	Property mapping	Blending	Thematic integration
Interpretation of noun–noun phrase pair	Property mapping (eg, "knife-shaped fork")	Hybrid (eg, "half fork, half knife")	Relation linking (eg, "knife and fork set")
Two-concept combining process	Metaphor method (eg, "white tomato")	Blending method (eg, "powdered ketchup")	Thematic relations method (eg, "moisture-keeping refrigerator")

As discussed, the two concepts (the noun set) can be utilized not only as base concepts but also as a noun–noun phrase pair for the purpose of interpretation. Therefore, we can discover the characteristics of the concept generation process by comparing these. Allow me to explain the results of an experiment carried out based on this idea (Nagai et al., 2009). This experiment was related to the design experiment outlined in Section 3.1.2. Experiment participants were given the task of interpreting the pairs of words "ship–guitar" and "desk–elevator" as a noun–noun phrase pair (the interpretation task) and the task of concept generation from these base concepts (the design task). These tasks were then compared. Each participant's response (in the case of the interpretation task, the response was their explanation of how they interpreted

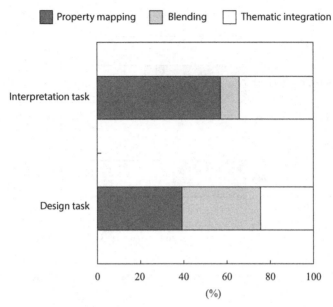

Figure 8.2 The difference between the methods of design and interpretation.

the noun–noun phrase pair, and in the design task, the response was the sketch and the sentences explaining it) was then classified into either "property mapping type (the metaphor method or property mapping)," "blending type (the blending method or hybrid)," or "thematic integration type (the thematic relations method, or relation linking)." The results indicated a greater proportion of blending type responses for the design task (concept generation) than the interpretation task responses (Fig. 8.2).

Moreover, the sets of words describing the design outcome (sketches and sentences) and the words explaining the interpretations[1] were classified based on whether they were related to the commonality between the base concepts, the alignable difference, or the nonalignable difference. The result was that a greater proportion of sets of words describing the design outcomes were classified as related to nonalignable difference than as explaining the interpretation (Fig. 8.3).

These results indicate that, compared with interpretation, concept generation (design) more often focuses on nonalignable difference, and uses the blending method.

[1] In the same way as the design task, participants are asked to list words describing their interpretation of the pair of words as a noun–noun phrase pair.

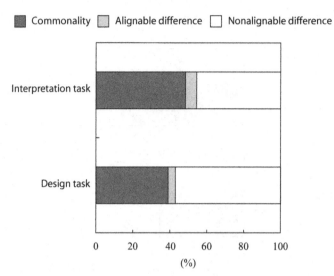

Figure 8.3 Difference between design and interpretation in terms of how similarity or dissimilarity is recognized.

8.2 FIRST-ORDER CONCEPT GENERATION AND HIGH-ORDER CONCEPT GENERATION

Now let us summarize the three methods of concept generation discussed in Chapters 4–6. These methods can be categorized into two groups (Taura and Nagai, 2013). The first is the metaphor method. In this method, a new concept (design idea) is generated to ensure a common property with the object used for comparison. The abstract concept created by this common property will be called a **first-order abstract concept**, and the generation of concepts based on first-order abstract concepts will henceforth be called **first-order concept generation**.[2] As shown in Fig. 8.4, the process of generating first-order abstract concepts can be illustrated as a process of creating a subset. For example, in creating <a tomato like snow>, the property "white" serves as a first-order abstract concept, and "snow" and "(the newly generated) white tomato" are elements of this subset. Another

---

[2] According to the similarity of metaphor and abduction discussed in Chapter 7, abduction would also fall under first-order concept generation. However, this consideration is based on the fact that abduction is understood within the framework of the formulation in that chapter. For instance, regarding the properties of abduction, Charles Peirce has said, "… although the possible explanations for our facts may be strictly innumerable, yet our mind will be able, in some finite number of guesses, to guess the sole true explanation of them" (Peirce and Burks, 1958). Based on this, abduction can be classified differently.

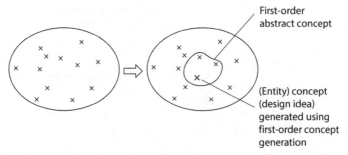

First-order
abstract concept

(Entity) concept
(design idea)
generated using
first-order concept
generation

Figure 8.4 First-order concept generation.

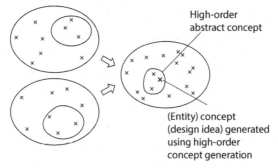

High-order
abstract concept

(Entity) concept
(design idea) generated
using high-order
concept generation

Figure 8.5 High-order concept generation.

example is the snap-off utility knife, where the property "snaps off" serves as a first-order abstract concept, and "a block of chocolate" and "the newly generated) snap-off utility knife" are elements of this subset.

The other group includes the blending method and the thematic relations method. These methods involve a manipulation that combines several abstract concepts. They are *high-order* methods in the sense that they "multiply" abstract concepts. Abstract concepts that are generated by performing such manipulative operations on abstract concepts will be called **high-order abstract concepts**, and the generation of new concepts based on high-order abstract concepts will henceforth be called **high-order concept generation**. As described in Chapter 5, the (nonalignable) difference between the base concepts plays an important role in high-order concept generation. High-order concept generation, as shown in Fig. 8.5, refers to finding new subsets from several existing subsets, and the subsets thus created are high-order abstract concepts.

As discussed in Chapters 4–6, first-order concept generation and high-order concept generation can both be expressed as processes of finding intersections of the abstract concepts created by base concepts (in first-order

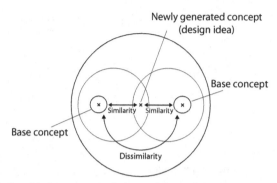

Figure 8.6 Relationship between similarity and dissimilarity in concept generation.

concept generation, one of the abstract concepts is a base concept itself, meaning that "addition" is performed). However, as shown in Fig. 8.6, the meaning of each changes depending on the type of relationship that is focused upon between the concepts. In first-order concept generation, the new concept (design idea) is generated focusing on the similarity between the new concept and the base concepts, whereas in high-order concept generation, the new concept (design idea) is generated focusing on the dissimilarity between the two base concepts.

Now, let us consider the difference between first-order and high-order concept generation from the perspective of the method of design. As discussed in Chapter 1, methods of design include analytical methods that are used when there is an explicit problem and synthetic methods that are used when there is no explicit problem.

First, we will examine first-order concept generation. First-order concept generation is often conducted in order to solve existing problems. Let us consider the example given in Chapter 4. The shape of the front carriage of the 500 Series Shinkansen was created to imitate a kingfisher in order to solve the pressure problems in the tunnel. Similarly, it is assumed that the process of inventing the snap-off utility knife involved thinking continuously about the existing problem of blades not staying sharp for long, thereby shortening the life of knives. While engaged in this thought process, the inventor happened to see a block of chocolate, which led to the idea of snapping off blades. In this way, first-order concept generation necessitates comprehensive understanding and analysis of existing problems. In this sense, it can be an analytical method.

High-order concept generation, on the other hand, involves not only comprehensive analysis of the base concepts and existing situations, but also the combining of properties extracted from the base concepts or the situations in which the products are used. Thus, we can say that high-order concept generation is a synthetic method. High-order concept generation can certainly be conducted without an explicit problem. Let us consider the example given in Chapters 5 and 6. "Powdered ketchup" was created by combining two different properties, "to fall lightly" and "acts as a flavoring" of "snow" and "tomato," respectively; however, this was not to solve any particular problem. Though not intended, a kind of problem may have been solved (eg, ketchup is difficult to use as a seasoning during meals), but this was only a result. The "moisture-keeping refrigerator" and "guitar that can be played on a ship" are similar examples. They resulted in solving some kind of problem; however, this problem was not the starting point for their invention. In this way, high-order concept generation can be engaged in even when the problem is not explicit.

This discussion provides a clue to answering the question that remained at the end of Chapters 4–6, "How are base concepts selected?"

First, if the problem is explicit, we know that we should select base concepts that would solve the problem (Sakaguchi et al., 2011). In the example of the snap-off utility knife, a block of chocolate was selected in order to solve the existing problem that blades did not stay sharp for long, thus shortening the life of knives, while in the example of the 500 Series Shinkansen, a kingfisher was selected in order to solve the existing problem of pressure in the tunnel.

If there is no explicit problem, we look for conditions generating more creative concepts, and rely on these to select our base concepts. Until now, the following have been revealed as such conditions.

- When the distance between the two base concepts is appropriately high, this will yield a design idea with high originality (Taura et al., 2005).
- When base concepts are such that many concepts can be associated with them, the design idea will be one with high originality (Nagai and Taura, 2006b).

8.3 THE SIGNIFICANCE OF PERFORMING HIGH-ORDER CONCEPT GENERATION

High-order concept generation can also be described as an extension of first-order concept generation. For example, the "powdered ketchup" idea in the "snow" and "tomato" example can also be generated by mapping a property of "snow" onto "ketchup." The idea of the "moisture-keeping refrigerator" can also be generated by mapping the property of "snow" onto "refrigerator." Thus, whether concept generation is high order or first order is not determined by the concept (design idea) that is generated as the result. The differences between the two are as follows.

The first difference, as outlined in the previous section, is whether an analytical method that presupposes the existence of a problem is used, or a synthetic method that allows concept generation even when there is no explicit problem is used.

The second difference is whether the generated concept is a subspecies of the base concepts or a new category in itself.

The third difference is whether similarity or dissimilarity is the crucial factor involved.

Incidentally, in actual design, the high-order concept generation process is not usually visible. This is because design without a clear objective or goal is not actually plausible. It is not realistic to apply high-order concept generation to "snow" and "tomato" in order to generate a practical idea, as we do not know which category of ideas will be generated until after the generation of ideas. When designing the utility knife, of course, we need the design to be a utility knife. So, what significance does high-order concept generation hold for design? For one, as it is a synthetic method, it has the potential to overcome the problem of the missing link in the pre-design phase. Moreover, it can help broaden the scope of thinking in concept generation. Concepts that are associated during concept generation, whether or not they directly constitute the final design idea, contribute indirectly by generating a wide range of alternative ideas. The relationship between creativity and the broadening of thoughts has already been discussed many times. The next section will briefly outline the main points of these discussions.

8.4 DIVERGENT THINKING IN CREATIVE DESIGN

It is said that thinking can be classified into "divergent thinking" and "convergent thinking." In this regard, an interesting explanation has been provided, which is outlined below (Weisberg, 1986). Divergent thinking is described as the "free-form" associative thinking that occurs when ideas are born, and convergent thinking as the normal logical thinking that occurs when ideas are applied to problems and the results evaluated. Then, the following test that evaluates the characteristics related to creative ability,[3] which emphasizes divergent thinking, is introduced.

---

[3] In this section, "creative" is used in the general sense (original, innovative, etc.), and not as defined in Chapter 1.

- Problem: list as many white, edible things as you can (3 min).
- Problem: list all the words that you can think of in response to "mother" (3 min).
- Problem: list all the uses that you can think of for a "brick" (3 min).

On the other hand, "…it was uniformly shown that performance on divergent thinking tests was unrelated to scientific creativity, as judged by other scientists in the same field. That is, the most creative scientists do not perform best on tasks involving divergent thinking, and those who perform better on the divergent thinking tests are not the most creative in their profession. Thus, divergent thinking tests do not measure the factors involved in scientific creativity" (Weisberg, 1986).

According to the discussion in Chapters 4–6, broader thinking is expected to generate outstanding concepts. However, is broad thinking really relevant to design? Below, the results of an experiment that sheds some light on this are outlined (Nagai and Taura, 2006a). Similar to the design experiment outlined in Section 3.1.2, the participants were required to generate new concepts (furniture in this instance) from two existing concepts. Specifically, the tasks they were given were "cat–hamster" and "cat–fish." In this design experiment, the participants were asked to say aloud the things they came up with or thought of during the task. This method, called protocol analysis, is widely used in the analysis of thinking. Nouns were then extracted from the participants' recorded utterances, and for each noun, the conceptual distances from the design target (furniture) and the base concepts (cat–hamster and cat–fish) were obtained. Conceptual distance, as shown in Fig. 8.7, was identified as the number of steps between the two concepts in the semantic network (concept dictionary) (in this study, EDR was used, EDR Concept Dictionary, 2005). In this example, the distance from "chair" to "furniture" was counted as 1 step, and that to "cat" as 12 steps.

Of the distances thus identified, if the X-axis indicates the distances from the base concepts and the Y-axis the distances from the target (furniture), each noun can be mapped in a 2D space. An example of the 2D thought space created based on the aforementioned process is shown in Figs. 8.8 and 8.9. Fig. 8.8 illustrates the 2D thought space for the design idea judged to have the highest originality.[4] This idea was generated from "cat–hamster" and named "chair that can be folded like an umbrella." It was described as, "a chair that

---

[4] Eight raters evaluated the design outcomes on a five-point scale (with 5 being the highest), the average of which was taken as the evaluated originality score. For details about the experiment, refer Nagai and Taura (2006a).

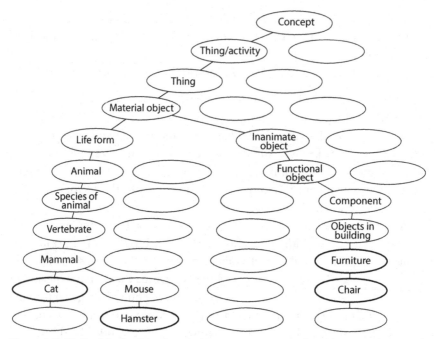

Figure 8.7 Method for finding distance between concepts using the semantic network (concept dictionary).

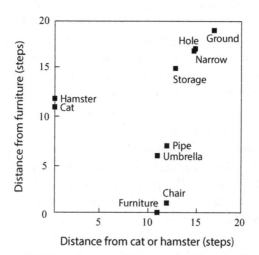

Figure 8.8 Example of 2D thought space (high evaluated originality score) (Nagai and Taura, 2006a).

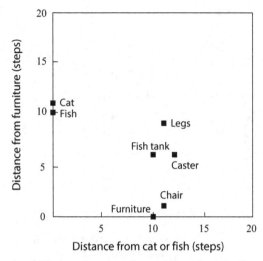

Figure 8.9 Example of 2D thought space (low evaluated originality score) (Nagai and Taura, 2006a).

can be folded using the mechanism of a folding umbrella and can be stored in a narrow space. It is possible to store it in an underground compartment after use." Fig. 8.9 illustrates the 2D thought space for the design idea judged to have the least originality. This idea was generated from "cat–fish" and named "fish tank with casters." It was described as, "there are legs like those of a chair attached to the bottom of the fish tank. Since they have casters, it is possible to move the tank easily." Comparing Figs. 8.8 and 8.9 reveals the broader expansion of thought employed in Fig. 8.8. In order to validate the relationship between the degree of expansion of thought and the originality of the design outcome, the distance from each of the points on the 2D thinking space (the spoken nouns) to the point of origin was determined, and the average of these values was calculated. This value is considered to represent the degree of expansion of thought. Fig. 8.10 shows the relationship between these values and the evaluated originality scores. A positive correlation was found between each ($0.05 < p < 0.10$). This result indicates that when devising an idea, greater expansion of thoughts will lead to ideas with high originality.

The results of this experiment seem to be inconsistent with the statement that "performance on divergent thinking tests was unrelated to scientific creativity, as judged by other scientists in the same field." This inconsistency seems to stem from a difference in the mechanism used for expansion of thought. The test that evaluates the characteristics related to creative ability seeks a larger number of associations, while these associations

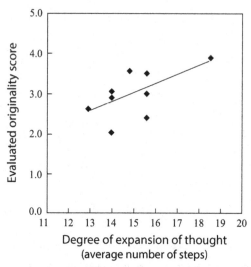

Figure 8.10 Relationship between expansions of thought and evaluated originality score (Nagai and Taura, 2006a).

actually bear no relation to each other. For example, in the test that asks about "white, edible things," an association is constantly made between "white" and "edible thing," meaning that no relationship is sought between each associated idea, and nor is such a relationship necessary.

Meanwhile, in the examples illustrated in Figs. 8.8 and 8.9, associations are made *sequentially*. In actuality, when the participants were asked about the relationships between each noun, they were able to explain more than 60% of the relationships. Here, let us focus on the simulation discussed in Section 3.1.2. In this simulation, the path between the base concepts and the set of words describing each design outcome is represented using the path between each set of two words in the semantic network (concept dictionary). Here, all of the words included in the path between the two words are incorporated into the simulation, meaning that the virtual concept generation process is modeled as a sequential path. The fact that the results of such a simulation explain the actual phenomenon (the evaluated originality score of the design outcome), that is, a correlation was found, indicates the possibility that the actual concept generation process has sequential properties. From this, we can assume that sequentially expanding associations, while relating concepts to each other, leads to the generation of concepts (design ideas) with high originality. We can thus assume that the associations made by creative scientists in actual research activities are not unrelated to each other; rather, they expand sequentially while being related.

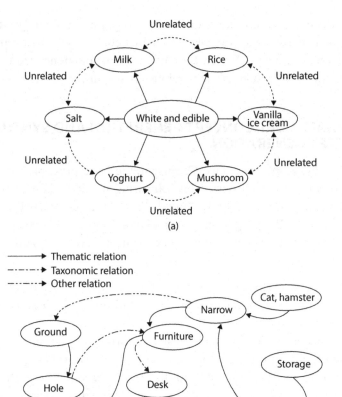

Figure 8.11 (a) Test that evaluates the characteristics related to creative ability; and (b) concept generation.

"Leaps of thinking" is a phrase that we hear within the context of creative thinking. The existence of such leaps seems to contradict our previous understanding. However, this contradiction can be resolved by the following hypothesis, "the creative thinking process is originally sequential, but it appears to be nonsequential because it is partially hidden within the inexplicit mind" (Taura et al., 2012).

To summarize, as Fig. 8.11 illustrates, the problem with the test that evaluates the characteristics related to creative ability is that because it seeks to find the expansion between unrelated associations, it differs greatly from the sequential associations involved in actual concept generation processes and scientific activities. We can conclude that this difference led to the inconsistency in the results of the previous discussions.

8.5 ANALYTICAL CONCEPT GENERATION AND SYNTHETIC CONCEPT GENERATION

Based on the discussions since Chapter 4, the basic principles of concept generation can be summarized by comparing it with a layered structure. This structure has six levels: the "motive level," the "perspective level," the "association level," the "concept manipulation level," the "concept generation level," and the "newness of generated concept level." They are discussed below in this order.

- The motive level is the level at which the type of motive that forms the basis of the concept generation is of significance. This level is classified into inner and outer motive.
- The perspective level is the level at which the point in question is what type of view is the basis for the concept generation being conducted. Similarity and dissimilarity form the basis of classification in this level.
- The association level is the level at which how the association expands is of significance. The important factors in this level are the expansion of thought and sequentiality discussed in this chapter, as well as the complexity discussed in Section 3.1.2.
- The concept manipulation level is the level at which the point in question is how concepts are manipulated. The focus is abstract ↔ concrete (abstracting and concretizing), and whole ↔ part (composing and decomposing) manipulation.
- The concept generation level is the level at which the point in question is how a concept is ultimately generated. This level is classified into concept generation using metaphor, and concept generation using blending or thematic relations.
- The newness of generated concept level is the level where classification is based on whether the generated concept is a subspecies of an existing category or a new category in itself.

Using the aforementioned categories, we can summarize the relationship between first-order concept generation and high-order concept generation as shown in Fig. 8.12.

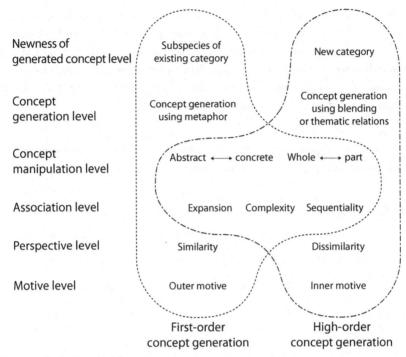

Figure 8.12 Basic principles of concept generation.

First-order concept generation is primarily based on outer motive, tends to focus on similarity, and expands associations sequentially and complexly, while repeating the manipulations of abstract ↔ concrete, and whole ↔ part, and generating concepts using metaphors. As a result, the generated concepts are subspecies of existing categories.

High-order concept generation is primarily based on inner motive, tends to focus on dissimilarity, expands associations sequentially and complexly, while repeating the manipulations of abstract ↔ concrete and whole ↔ part, and generating concepts using blending and thematic relations. As a result, the generated concepts are themselves new categories.

Here, we refer to concept generation engaged when problems are explicit and social motive is fully matured as **analytical concept generation**, and that it is not like **synthetic concept generation**. Based on this classification, first-order concept generation is one method for the former, and high-order concept generation is one method for the latter. Moreover, inferring latent fields and functions as discussed in Chapter 3 is an effective method for synthetic concept generation.

Part 2 of this book (Chapters 4–7) has discussed methods for concept generation while referring to existing concepts (base concepts). However, new concepts may also be generated directly or instinctively with absolutely no relation to existing concepts. The basic principles and methodology of this kind of concept generation are extremely difficult to grasp, and thus are outside the scope of the discussion in this book.

Task

Crows also appear to have ingenuity. Can we say that crows design? If so, based on what principles can it be considered that they design?

REFERENCES

EDR Concept Dictionary 2005, EDR electronic dictionary (CD-ROM). National Institute of Information and Communications Technology, CPD-V030.

Nagai, Y., Taura, T., Mukai, F., 2009. Concept blending and dissimilarity: factors for creative concept generation process. Des. Stud. 30, 648–675.

Nagai, Y., Taura, T., 2006a. Formal description of the concept-synthesizing process for creative design. In: Gero, J.S. (Ed.), Design Computing and Cognition. Springer, London.

Nagai, Y. Taura, T., 2006b. Role of action concepts in the creative design process, in Proceedings of the International Design Research Symposium, Kim, Y.S. (Ed.), Seoul, pp. 257–267.

Peirce, C.S., Burks, A.W. (Eds.), 1958. Collected Papers of Charles Sanders Peirce, vol. VII, Harvard University Press, Cambridge, MA, 7.219.

Sakaguchi, S., Tsumaya, A., Yamamoto, E., Taura, T., 2011. A method for selecting base functions for function blending in order to design functions. Proceedings of Eighteenth International Conference on Engineering Design, vol. 2. Copenhagen, pp. 73–86.

Taura, T., Nagai, Y., 2013. A systematized theory of creative concept generation in design—first-odder and high-order concept generation. Res. Eng. Des. 24 (2), 185–199.

Taura, T., Nagai, Y., Tanaka, S., 2005. Design space blending—a key for creative design. Proceedings of ICED: International Conference on Engineering Design, Melbourne.

Taura, T., Yamamoto, E., Fasiha, M.Y.N., Goka, M., Mukai, F., Nagai, Y., Nakashima, H., 2012. Constructive simulation of creative concept generation process in design—a research method for difficult-to-observe design-thinking process. J. Eng. Des. 23, 297–321.

Weisberg, R.W., 1986. Creativity: Genius and Other Myths. WH Freeman and Co., New York.

Wisniewski, E.J., 1996. Construal and similarity in conceptual combination. J. Mem. Lang. 35, 434–453.

PART 3

Theory and Methodology of Conceptual Design

CHAPTER 9

Methodology of Conceptual Design

Contents

9.1 AN OVERVIEW OF DESIGN METHODOLOGY

This chapter will introduce the design methodology of Gerhard Pahl and Wolfgang Beitz, which is highly regarded within engineering design methodology. A very simple outline of selected content explained in the work of Pahl and Beitz (1995) is provided below with some modifications.

As shown in Fig. 9.1, machines can be described as systems where inputs and outputs cross the system boundary, and inputs are converted to outputs. The following three types of inputs and outputs are discussed here.

- Energy: mechanical, thermal, electrical, chemical, optical, nuclear, etc.; also force, current, heat, etc.
- Material: gas, liquid, solid, dust, etc.; also raw material, test sample, workpiece, etc.; also end product, component, etc.
- Signals: magnitude, display, control impulse, data, information, etc.

In every type of conversion, we must take into consideration both quality and quantity.

As shown in Fig. 9.2, design proceeds in the order of **clarification of the task**, **conceptual design**, **embodiment design**, and **detail design**. Each stage is briefly explained below.

1. Clarification of the task: at this stage, specifications (requirements or constraints) required for design solutions are listed, and a detailed specifications document is elaborated. It must be determined whether each specification is a demand or a wish. "Demands" are requirements or

Creative Design Engineering
http://dx.doi.org/10.1016/B978-0-12-804226-7.00009-0

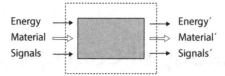

Figure 9.1 Description of a machine. *(Modified from Pahl and Beitz (1995)).*

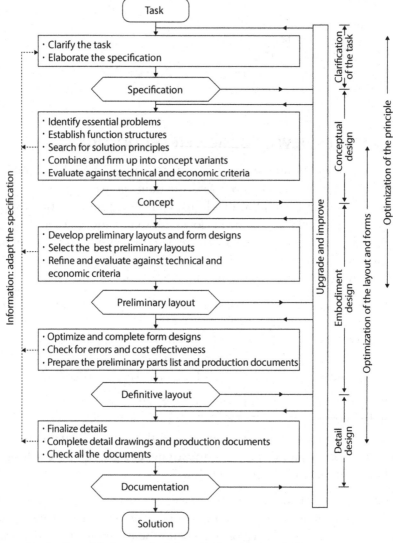

Figure 9.2 Outline of design methodology (Pahl and Beitz, 1995).

Table 9.1 Items that should be considered as specifications (Pahl and Beitz, 1995)

Main headings	Examples
Geometry	Size, height, breadth, length, diameter, space requirement, number, arrangement, connection, and extension.
Kinematics	Type of motion, direction of motion, velocity, and acceleration.
Forces	Direction of force, magnitude of force, frequency, weight, load, deformation, stiffness, elasticity, inertia forces, and resonance.
Energy	Output, efficiency, loss, friction, ventilation, state, pressure, temperature, heating, cooling, supply, storage, capacity, and conversion.
Materials	Flow and transport of materials. Physical and chemical properties of the initial and final product, auxiliary materials, and prescribed materials (food regulations. etc.).
Signals	Inputs and outputs, form, display, and control equipment.
Safety	Direct protection systems, operational and environmental safety.
Ergonomics	Man–machine relationship, type of operation, operating height, clearness of layout, sitting comfort, lighting, and shape compatibility.
Production	Factory limitations, maximum possible dimensions, preferred production methods, means of production, achievable quality, tolerances, and wastage.
Quality control	Possibilities of testing and measuring, application of special regulations and standards.
Assembly	Special regulations, installation, siting, and foundations.
Transport	Limitations due to lifting gear, clearance, means of transport (height and weight), nature and conditions of dispatch.
Operation	Quietness, wear, special uses, marketing area, destination (eg, sulfurous atmosphere, tropical conditions).
Maintenance	Servicing intervals (if any), inspection, exchange and repair, painting, and cleaning.
Costs	Maximum permissible manufacturing costs, cost of tools, investment, and depreciation.
Schedules	End date of development, project planning and control, and delivery date.

constraints that must be met under all circumstances. "Wishes" are requirements or constraints that should be taken into consideration wherever possible. A list of major specification items is provided in Table 9.1. Such requirements or constraints are summarized in a specifications document, as can be seen in Fig. 9.3.

2. Conceptual design: at this stage, the overall structures and mechanisms are devised. In this methodology, function structures are first established,

Issued: xx/20xx

Changes	D W	Requirements	
		Specification for **Sub-task: assemble cartons**	Page: 1
		Requirements	Resp.

	W	Assemble and glue 15 cartons/min
		Size of bought-out sections
		Alternatives 500 x 500 mm
		                        400 x 400 mm
		                        450 x 450 mm (only 10%)
		Probably tolerance: ± 1 mm
		Cardboard sections are fed in manually. However, allow for automatic feed-in in due course
		(Project conference minutes 16/70)
		Assembled cartons placed on conveyer belt lying on their base. Height of conveyor belt above floor level: 300 mm	

	W	Cartons capable of being removed in any of the three directions 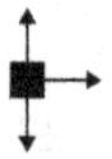	
		Counter required for counting the assembled cartons	
	W	Machine must be quickly movable without further adjustment Gluing: On leaving the machine, the glue must have set, and the cartons must be capable of bearing the full load	
	W	Working principle must allow increase in output to 30 cartons/min with automatic feed mechanism Maximum production costs xxxx (project conference minutes 20/70) Schedule: End of development xx/xx/20xx Planned delivery date xx/xx/20xx	
		Replaces issue of	

Figure 9.3 Example of a specifications document (carton assembly machine). *(Modified from Pahl and Beitz (1995)).*

before searching for suitable solution principles for the realization of each element that makes up the structured functions. These solution principles are then combined and firmed up into concept variants. This will be discussed in detail in the next sections.

3. Embodiment design: at this stage, the layout and forms are determined. Latest by this stage, technical and economic feasibility are checked.

4. Detail design: at this stage, all the dimensions and materials required for each component are specified, and technical and economic feasibility are rechecked, before drawings and other production documents are produced.

9.2 FUNCTION DECOMPOSITION

As shown in Fig. 9.4, the following order of stages should be followed in conceptual design: problem abstraction, establishing function structures, searching for solution principles, combining solution principles into concept variants, selection and confirmation of suitable combinations, and the evaluation of the created concept variants.

At the stage of abstracting the problem and identifying its essence, the crux of the task is clarified, ignoring what is particular or incidental and emphasizing what is general and essential. The specific procedure for this is outlined below.

Step 1: eliminate personal preferences.

Step 2: omit requirements or constraints that have no direct bearing on the function and the essential constraints.

Step 3: transform quantitative data into qualitative data and reduce the data to essential statements.

Step 4: generalize the results of the previous step.

Step 5: formulate the problem in solution-neutral terms.

At the stage of establishing function structures, those functions manifested by the product as a whole (these are called whole functions[1]—refer Chapter 4) are decomposed into their subordinate functions (these are called partial functions—refer Chapter 4). Partial functions[2] are functions that are manifested by parts of a product, and the whole function is fulfilled through a set of these. In this book, finding the partial function set (this will be called **function structure**) from the whole function will be called **function decomposition**. The difficulty of function decomposition differs depending on whether the design is an original, adaptive, or variant design (based on the categories given in Section 2.3.2). In the case of original design, neither the partial functions nor the relationships between them are

[1] This corresponds to "overall function" in Pahl and Beitz (1995).

[2] This corresponds to "sub-function" in Pahl and Beitz (1995). As will be discussed in Chapter 10, this book will classify functions obtained through decomposition of the whole function into two types. The first are "partial functions," and the second, defined in Section 10.2.2, will be called "sub functions."

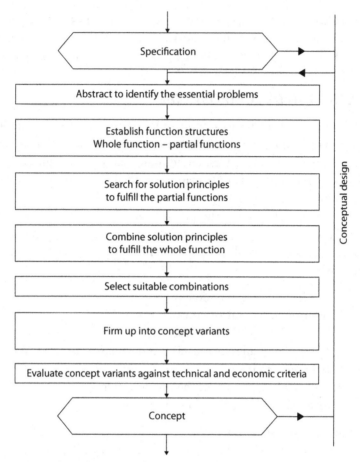

Figure 9.4 Procedure of conceptual design. *(Modified from Pahl and Beitz (1995)).*

generally known. Therefore, an optimum function structure must be newly established, and this constitutes some of the most important steps of the conceptual design phase. In the case of adaptive and variant designs, on the other hand, the whole structure of the product is often already known, and the function structure is comparatively easy to establish.

The following methods can be used for function decomposition.

The first method is analyzing the logical relationships between partial functions. For example, when relationships such as "if partial function A is present, then partial function B can come into effect," or logical relationships such as AND functions and OR functions exist, a function structure that relies on these can be established. Fig. 9.5 shows a car door catch mechanism. The logical relationship in this case is an AND

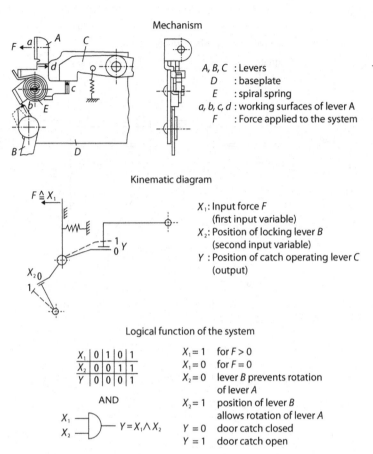

Figure 9.5 Logical relationships in the catch mechanism of a car door (Pahl and Beitz, 1995).

function. Operating lever C can only be activated by the input force F acting on the lever A if the locking lever B is at "1."

The second method is based on describing a machine as a system that converts energy-, material-, and signal-related inputs to outputs. When the flows of energy, material, and signals within a machine are clearly described, function decomposition can be performed by dividing these flows (Fig. 9.6).

Note that these methods are applicable only under certain relationships between whole and partial functions. This will be described in detail in Chapter 10.

The function structure of a tensile testing machine created based on the previous discussion is shown in Figs. 9.7 and 9.8. These describe the main essential partial functions that are directly related to the whole function (Fig. 9.7) and the detailed function structure including the auxiliary partial functions (Fig. 9.8).

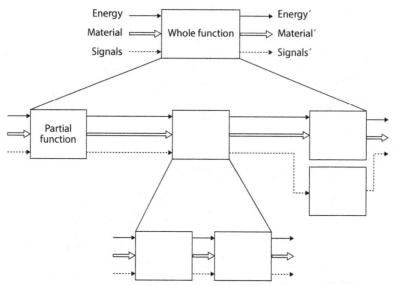

Figure 9.6 Function decomposition based on the flow of energy, material and signals. *(Modified from Pahl and Beitz (1995)).*

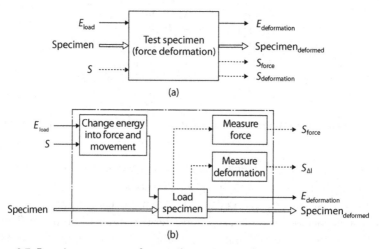

Figure 9.7 Function structure of a tensile testing machine (a) whole function; and (b) main essential partial functions (Pahl and Beitz, 1995).

9.3 SEARCHING FOR SOLUTION PRINCIPLES

In this stage, the mechanisms needed to fulfill the partial functions identified in the previous section are searched for. In this book, these are called solution principles (Section 2.3.2). Pahl and Beitz (1995) introduce

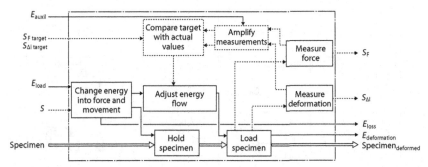

Figure 9.8 Complete function structure of a tensile testing machine (Pahl and Beitz, 1995).

the method of analogy from natural systems; idea generating methods such as brainstorming, the Delphi method, and synectics; the method of using physical effect; and the method of utilizing design catalogs. Of these, this book will outline the "method of using physical effect." The method of analogy from natural systems is similar to the contents of the discussion in Section 4.3.1 and the remaining methods will not be discussed here, as they are general tools applicable not only to design but to a range of thinking processes.

In mechanical design, solution principle behavior always conforms to some kind of "physical effect." Physical effects can be described by means of the physical laws governing physical phenomena. For example, the friction effect is described by Coulomb's law, the lever effect by the lever law, and the expansion effect by the expansion law. Each partial function is fulfilled by various physical effects. For example, the partial function of amplification can be fulfilled by lever effect, wedge effect, electromagnetic effect, hydraulic effect, etc. Thus, for each partial function, a number of physical effects can be sought. The procedure for searching for solution principles using physical effects is as follows:

Step 1: seek physical effect (A), which enables fulfillment of partial function (B).
Step 2: using the physical effect sought in Step 1, create the description "partial function (B) is fulfilled by physical effect (A)" ("physical principle"). Step 3: from the description obtained in Step 2, devise a specific mechanism to act as the solution principle.

Fig. 9.9 shows an example of solution principles searched for using this method.

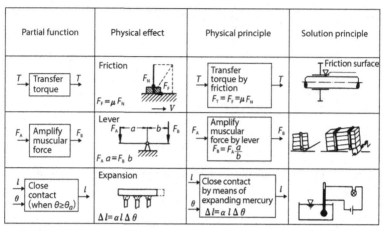

Figure 9.9 Method of searching for solution principles based on physical effects. *(Modified from Pahl and Beitz (1995)).*

9.4 FIRMING UP INTO CONCEPT VARIANTS BY COMBINING SOLUTION PRINCIPLES

The solution principles searched for in the previous section are combined to firm them up into "concept variants." These are obtained by selecting one solution principle for each partial function and combining solution principles. Let us consider Fig. 9.10. The figure shows an example of selecting the solution principles outlined in bold from each of the partial functions <lift>, <sift>, <separate leaves>, <separate stones>, <sort potatoes>, and <collect>. Note that when selecting solution principles, we must be careful to maintain physical and geometrical connectivity in their combinations. It is also advisable to find a number of combinations and create multiple concept variants.

9.5 EVALUATING CONCEPT VARIANTS

At this stage, the firmed–up concept variants are evaluated against technical and economic criteria. Table 9.2 illustrates a checklist for evaluation. Evaluation criteria are determined based on such a checklist, and a quantitative evaluation is undertaken. A feasibility study is performed, and cost, strength, and output values are calculated. Note that when conducting an evaluation, it is important to estimate the level of uncertainty.

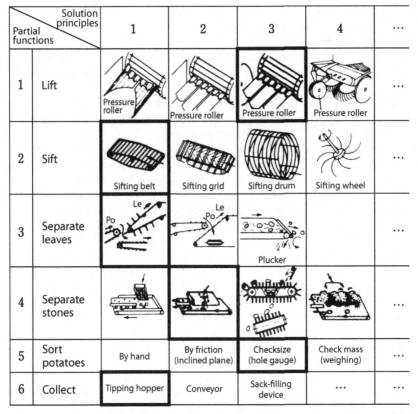

Partial functions	Solution principles	1	2	3	4	...
1	Lift	Pressure roller	Pressure roller	Pressure roller	Pressure roller	...
2	Sift	Sifting belt	Sifting grid	Sifting drum	Sifting wheel	...
3	Separate leaves			Plucker		...
4	Separate stones					...
5	Sort potatoes	By hand	By friction (inclined plane)	Checksize (hole gauge)	Check mass (weighing)	...
6	Collect	Tipping hopper	Conveyor	Sack-filling device	...	...

Figure 9.10 Example of combining solution principles (potato harvesting machine). *(Modified from Pahl and Beitz (1995)).*

Task

Follow these steps to generate concept variants for the new concepts of products generated by using the methods described in Chapters 4–6.

Step 1: create a specifications document.

Step 2: based on the specifications document, establish a function structure, similar to the one in Fig. 9.8.

Step 3: search for the solution principles for the important partial functions, similar to Fig. 9.9, and summarize them into a list, as indicated in Fig. 9.10.

Step 4: based on the list created in Step 3, combine solution principles and determine the concept variants. Make sketches of the determined concept variants.

Step 5: evaluate the concept variants determined in Step 4 using calculations or experiments.

Table 9.2 Checklist for evaluating concept variants

Main headings	Examples
Function	Do the essential main and auxiliary functions follow necessarily from the selected solution principle or concept variant?
Working principle	Is the selected principle or function simple and clear-cut, expected to produce an adequate effect, and hampered by few disturbing factors?
Embodiment	Does this have small number of components, and is there low complexity and low space requirement? Are there no special problems with layout or form design?
Safety	Are there no additional safety measures needed? Are industrial and environmental safety guaranteed?
Ergonomics	Is there a satisfactory man–machine relationship? Is there no strain or impairment of health? Is the design form good?
Production	Can this be produced by a small number of established production methods? Is the equipment inexpensive? Is this made of a small number of simple components?
Quality control	Are there few tests and checks needed? Are procedures simple and reliable?
Assembly	Can this be done easily, conveniently, and quickly? Are no special aids needed?
Transport	Can normal means of transport be used? Are there no risks?
Operation	Is operation simple? Is there a long service life? Is low wear expected? Is handling easy and simple?
Maintenance	Are upkeep and cleaning minimal and simple? Is inspection easy? Is repair easy?
Costs	Are there no special running or other associated costs? Are there no scheduling risks?

Modified from Pahl and Beitz (1995).

REFERENCE

Pahl, G., Beitz, W., 1995. Engineering Design: A Systematic Approach. Springer, Berlin.

CHAPTER 10

Basic Principles of Conceptual Design

Contents

10.1 THEORETICAL FRAMEWORK OF CONCEPTUAL DESIGN

Let us simplify the procedure for conceptual design explained in the previous chapter as follows: whole function → partial function → solution principle → concept variant. We now depict it on a 2D plane as shown in Fig. 10.1.

In this figure, we can see that expressions related to the whole of the product (whole functions and concept variants) are located in the top half, and expressions related to parts of the product (partial function and solution principles) are located in the bottom half. Meanwhile, the left side of the diagram (whole functions and partial functions) indicates the expressions that are presented in abstract terms, and the right side of the diagram (concept variants and solution principles) indicates the expressions that are presented in a concrete form. From this, we can see that this 2D plane is composed of the abstract–concrete horizontal axis and the whole–part vertical axis.

This shows that conceptual design is a combination of the operation processes, abstract ↔ concrete (abstracting and concreting), and whole ↔ part (composing and decomposing). Note that, as discussed in Chapters 4–8, the operations of abstract ↔ concrete and whole ↔ part, are a key aspect of the concept generation process. The abstract ↔ concrete and whole ↔ part processes also have a strong connection to the fundamental operations of the rhetorical techniques, synecdoche, and metonym. This illustrates how the concept operation performed in concept generation and conceptual design is an ordinary and nonparticular activity.

Creative Design Engineering
http://dx.doi.org/10.1016/B978-0-12-804226-7.00010-7
119

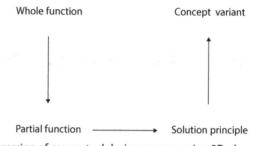

Figure 10.1 Expression of conceptual design process using 2D plane.

10.2 CHARACTERISTICS AND METHODOLOGY OF FUNCTION DECOMPOSITION

10.2.1 Characteristics of Function Decomposition

Let us consider the importance of function decomposition. Pahl and Beitz have stated that "function structures are intended to facilitate the discovery of solutions—they are not ends in themselves" (Pahl and Beitz, 1995). They also state, "the search for, and establishment of, an optimum function structure constitute some of the most important steps of the conceptual design phase." Design requires nothing more than the devising of concept variants from demands or wishes (whole functions), and therefore, decomposing these into partial functions and searching for solution principles is a long way of going about doing things. Why then must we take this route? We deliberate on the reasons for this discussion below.

First, it makes searching for solution principles easier. Some partial functions allow the use of existing components of the product (as they are or with some modifications) as solution principles. Alternatively, it is easier to devise new solution principles when the scope of the functions is limited. Thus, rather than devising solution principles directly from demands or wishes (whole functions), it is simpler to decompose these and consider at the level of partial functions here.

Second, the process of decomposing whole functions into partial ones and search for solution principles has a role in explaining design. Not only those directly involved in the manufacturing sector, but also others hope for an all-encompassing explanation for design. There are two ways of explaining design: explanation of the product and explanation of the process. The former involves stating the structure and functions of a designed product. For example, the explanation of the design of a car would involve describing how the driving mechanism works and how well it performs. The latter involves stating the history and intention behind the product. This is particularly convincing when the process of function decomposition is employed. For example, the "potato harvester" illustrated in Chapter 9 (Figure 9.10) can be explained as follows. In order to satisfy the demand or wish "to harvest potatoes" (whole function), the partial functions <lift>, <sift>, <separate leaves>, <separate stones>, <sort potatoes>, and <collect> (partial functions) are identified. Several possible methods (solution principles) can be used for fulfilling each of these functions (partial functions). For example, the partial function <sift> can be fulfilled using a "sifting belt," "sifting grid," "sifting drum," or "sifting wheel." In this design, "sifting belt" was chosen. In this way, stating the process of conceptual design from the perspective of function decomposition provides an explanation for the design.

Third, using function decomposition makes it easier to change the design. In design, changes are constantly being made. Although some of these changes arise from following the design process inappropriately, most of them occur because design change is an inevitable characteristic of design. As design involves devising products that do not exist yet, whether or not the devised product satisfies the relevant demands or wishes can only be evaluated after the product has materialized. Until then, it is nothing more than a *temporary fix*. Therefore, it is constantly subject to revision and alteration. When the product is revised or altered, if function decomposition has not been performed, the entire concept variant needs to be reconsidered. This would mean repeating almost exactly the steps involved in the first design. On the other hand, if function decomposition has been performed suitably, revision or alteration can be applied only to the corresponding partial function. Furthermore, if multiple solution principles have already been devised for each partial function, it is easy to find an alternative solution principle for revision or alteration. This also allows us to make explicit which part (solution principle) has been changed in what way, meaning that the history and intention of the change can be easily shared between those involved.

For all these reasons, function decomposition is a useful process. Function decomposition plays a particularly vital role in original design.

Meanwhile, function decomposition is not an ordinary process but indeed a particular one. This will be explained below. In mechanical design, the assembly of components can be explicitly expressed on paper. In other words, if we add up component attributes (shape, movement, etc.), then the attributes of the assembled product can be uniquely obtained. For this reason, the relationship in which the whole (the form of the assembled product) is uniquely obtained from the parts (components) will be called a

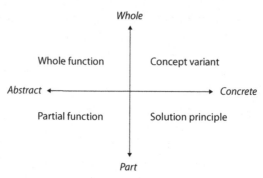

Figure 10.2 Structure of conceptual design.

linear whole–part relationship. The relationship between the whole and parts indicated on the right side of Fig. 10.2 is a linear whole–part relationship.

So what is the relationship between whole and parts on the left side of the same diagram? As discussed in Chapter 5, functions are considered to be manifested from a number of entity attributes. Thus, functions are those that are abstracted (extracted) from attributes. Specifically, partial functions are those abstracted from solution principles,[1] and whole functions are those that are abstracted from concept variants. If we view things in this manner, the relationship between whole functions and partial functions can be described as, "the whole–part relationship obtained by abstracting the attributes of both the whole and the parts in the linear whole–part relationship." So, what kind of "whole–part relationship" is obtained from "abstracting the attributes of both the whole and the parts in the linear whole–part relationship"?

Let us look at Fig. 10.3 for an example. This is a combination of gears. If we combine a gear with a gear ratio of 3.0 with another that has a gear ratio of 0.1, the combined gears will form a reduction gear with a ratio of 0.3. If we abstract the attributes (gear ratios) of each of the gears in this combination, the expressed partial functions obtained are <increase speed> and <reduce speed>. On the other hand, the whole function is <reduce speed>. That is, the relationship between the whole function and the partial functions ceases to be a "linear whole–part relationship." We cannot uniquely determine whether the whole function of the combined gears will be <increase speed> or <reduce speed> simply from the knowledge

---

[1] Since solution principles are searched from partial functions in the design process, the expression "to abstract solution principles" seems to defy the common notion of the process of the temporal progression of design. However, if we focus solely on the natures of both of these, it can be interpreted as this statement.

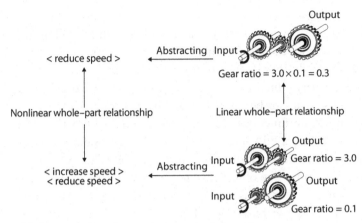

Figure 10.3 Linear and nonlinear whole–part relationships.

that the partial functions are <increase speed> and <reduce speed>, respectively. In this book, such a relationship, wherein the form of the whole cannot be uniquely determined from the forms of the parts will be called a **nonlinear whole–part relationship**. For functions that exhibit a nonlinear whole–part relationship, function decomposition becomes extremely difficult. The methods of decomposition outlined in Section 9.2 are effective, precisely because they are linear whole–part relationships based on function descriptions relying on input/output relationships. In case of highly abstract functions, we must think of a different method to use, which will be discussed in Section 10.2.3.

10.2.2 Vertical and Horizontal Decomposition of Functions

Let us consider functions in further detail. As discussed in Chapter 9, the function manifested by the product as a whole is the whole function. The whole function is composed of a set of multiple, more detailed functions. In this book, the elementary functions that thus comprise a function (A) will be called **sub functions**[2] of function (A). Note that what are referred to as sub functions here are the externally observable elementary functions of the product and necessary to *constitute the whole* of function (A). Thus, the objects (targets) of the verb (action) of sub functions are the same as those of function (A). For example, the whole function of a washing machine

---

[2] In Pahl and Beitz (1995), this term corresponds to "sub-function", but in this book, the functions that whole functions can be decomposed into are classified into two categories. The first are called "partial functions" and the second, which will be defined here, are called "sub functions."

<wash clothing> is composed of the three sub functions <wash clothing with detergent>, <rinse clothing>, and <drain clothing>. Here, the object of the verb of both the whole function and the sub functions is "clothing."

Meanwhile, this book refers to functions that are manifested by *parts* of products as partial functions. There are two types of partial functions. The first are those that can be observed externally as one element of a whole function (ie, a sub function). For example, the partial function of a plastic bottle cap, <act as a lid>, can be observed from outside the bottle as one element of a whole function (ie, a sub function). The second are partial functions that cannot be observed externally. For example, the partial function of a car engine, <turns shaft>, is manifested internally in the product and cannot be observed from outside the car.

Based on the aforementioned categories of partial functions, function decomposition can be classified into two different types (Yamamoto et al., 2010). The first is the decomposition of whole functions into partial functions that can be observed externally. This will be called **vertical decomposition of functions**.[3] In the vertical decomposition of functions, whole functions are first decomposed into sub functions, after which the partial functions that manifest those sub functions are identified. Here, the object of the whole function and the partial function are the same. For example, if we return to the example of the washing machine, its three sub functions <wash clothing with detergent>, <rinse clothing>, and <drain clothing> can be expanded (for a twin–tub washing machine) as <washing tub washes clothing with detergent >, <washing tub rinses clothing>, and <spin tub drains clothing>. Fig. 10.4 is a schematic representation of the vertical decomposition of functions.

The second type of function decomposition is the decomposition of whole functions or sub functions into partial functions that are manifested internally in the product (which cannot be observed externally). This will be called **horizontal decomposition of functions**.[4] In the horizontal decomposition of functions, partial functions that are causally connected are sought. For example, the whole function of cars <transport people and things> can be decomposed into the causally connected partial functions, <engine turns shaft>, <shaft turns tires>, <tires move vehicle body>, and <vehicle body carries people and things>. In the horizontal decomposition of functions, the object of a partial function becomes the subject of the

---

[3] In Yamamoto et al. (2010), this term corresponds to "decomposition-based dividing."

[4] In Yamamoto et al. (2010), this term corresponds to "causal-connection-based dividing."

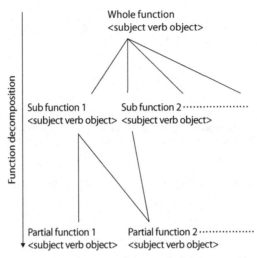

Figure 10.4 A schematic representation of the vertical decomposition of functions.

next partial function in the causal chain. Note that the final partial function can be observed externally as one element of the whole function (ie, as a sub function) and the object of the final partial function is the same as the object of the whole function or sub function. Fig. 10.5 represents the horizontal decomposition of functions.

10.2.3 Function Decomposition Using a Hierarchical Lexical Database

This section will explain a method for the decomposition of functions with a high level of abstraction using hierarchical word relationships (Yamamoto et al. (2010); Taura and Nagai (2012)).

When functions are described as a verb–noun set, that is, <verb object> or <subject verb object>, function decomposition can be performed by looking for suggestions from the decomposition of each verb and noun (subject and object). If, for instance, a database were prepared that stored the whole–part hierarchical word relationships between verbs and those between nouns, sub or partial functions could be obtained by searching for nouns or verbs that fall within the subordinate category of each noun or verb in a given whole function. For example, for the whole function <wash clothing>, the database would allow us to find "rinse" as a verb below the verb "wash."

Incidentally, semantic networks (concept dictionaries) that have been built thus far mainly include abstract–concrete relationships between words.

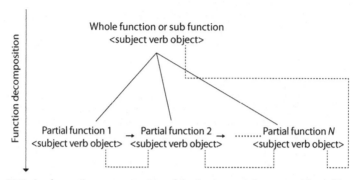

Figure 10.5 A schematic representation of the horizontal decomposition of functions.

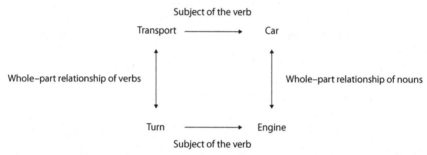

Figure 10.6 Whole–part relationships of verbs.

Thus, information regarding whole–part relationships must be newly collected. When doing so, whole–part relationships between verbs must be considered particularly carefully. This is because for the whole–part relationship between verbs in function decomposition, the nouns that are the subjects of the verbs must be the same as or belong to the whole–part relationship (Fig. 10.6).

Furthermore, when there are causal connections between multiple partial functions, lateral relationships must be taken into account. To do so, for a certain whole function, the two types of descriptions <subject verb> and <verb object> belonging to the subordinate category of the whole function should first be extracted from the text, after which the two descriptions should be connected using the verbs they share. Finally, those that satisfy the conditions for horizontal function decomposition should be extracted from the connected <subject verb object> description. An example is the case of the whole function <fan emits air> for which the partial functions

<motor turns impeller> and <impeller compresses air> are identified. In this example, the two partial functions are causally connected, and between the whole and partial functions, whole–part relationships between each of the subjects as well as whole–part relationships between verbs based on these subjects also exist.

10.3 FUNCTION DECOMPOSITION MATRIX

Let us consider the relationships between whole functions and partial functions in more detail. A matrix that shows whether or not relationships exist between partial functions and sub functions that comprise whole function will be called a function decomposition matrix. Cases where a partial function contributes to the manifestation of a sub function are indicated using an × in the matrix. When functions are decomposed vertically, partial functions directly manifest sub functions, but when functions are decomposed horizontally, they indirectly contribute to sub functions as internal functions. In a function decomposition matrix, an × is used to indicate a relationship between a sub function and partial function, regardless of whether it is direct or indirect.

As illustrated in Fig. 10.7, different patterns can be obtained from a matrix where each row is a partial function and each column a sub function, depending on the arrangement.

The first pattern (a) is that where several blocks line up diagonally. This will be called "fully segmented-type function decomposition."

The second pattern (b) is that where several blocks line up horizontally. This will be called "shared-type function decomposition."

The third pattern (c) is that where several blocks line up vertically. This will be called "adjusted-type function decomposition."

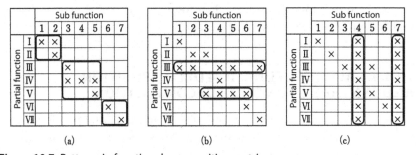

Figure 10.7 Patterns in function decomposition matrices.

In fully segmented-type function decomposition, design can be conducted separately within each block of partial functions. Furthermore, where those blocks are independent of each other (functions do not affect each other), "modular design" can be engaged in.

In shared-type function decomposition, the solution principle blocks lined up horizontally become the parts shared among some sub functions in the design solution. For example, car batteries are shared power sources contributing to many sub functions such as lighting and audio.

In adjusted-type function decomposition, because a number of partial functions are related to one sub function, these need to be adjusted. For example, low vibration in cars is related to the partial functions of several parts, including the engine, suspension, and structure of the car body.

Until now, in the field of design, modular design has received attention, and the trend has been to decompose whole functions into several modules of partial functions that are independent of each other. This is because it allows design and manufacturing to be conducted efficiently. If whole functions can be decomposed into independent modules, the design and manufacture of each module can be performed independently. This enables division of labor for design and manufacturing. Furthermore, as each module becomes a black box, quality management becomes easier and product reliability increases.

However, for both natural and man-made objects, the relationship between partial and whole functions is generally complex. Modularization runs the risk of oversimplifying the relationship between the whole and parts, which was originally complex. By itself, it narrows the scope of products that can be designed. Functions are also concepts that form links between users and products. Some functions are even discovered through the interactions between users and products. If we widen our perspective to include not only functions based on physical characteristics but also those based on emotional feelings, a large variety of functions can link users and products. Taking this discussion into account, adjusted-type function decomposition is thought to hold clues to the realization of designs that are highly aligned with the desires of society.

10.4 ANALYTICAL FUNCTION DECOMPOSITION AND SYNTHETIC FUNCTION DECOMPOSITION

In adaptive design and variant design, having an existing product as a reference point means that the whole function is often described in concrete terms using input and output relationships. In such cases, the relationship between whole functions and partial functions is a linear whole–part relationship, and the analysis of these whole functions allows for performing function decomposition. Such function decomposition will be called **analytical function decomposition**. For example, the method introduced in Section 9.2 is an effective method of analytical function decomposition. Meanwhile, in original design, whole functions are often only described in abstract terms. In such cases, the relationship between whole functions and partial functions is a nonlinear whole-part relationship, and a function structure needs to be established by mixing fragmentary knowledge. Such function decomposition will be called **synthetic function decomposition**. For example, the method introduced in Section 10.2.3 is an effective method of synthetic function decomposition.

In this chapter, we have discussed the methods of decomposing whole functions or sub functions into their partial functions. It should be noted that partial functions or their sub functions could be further decomposed into their subordinate functions using the very same methods.

Task

First, try to design a card shuffler (a device that arranges stacked cards in a random order) without performing function decomposition. Next, try it after performing function decomposition.

REFERENCES

Pahl, G., Beitz, W., 1995. Engineering Design: A Systematic Approach. Springer, Berlin.
Taura, T., Nagai, Y., 2012. Concept Generation for Design Creativity: A Systematized Theory and Methodology. Springer, Berlin, pp. 133–149.
Yamamoto, E., Taura, T., Ohashi, S., Yamamoto, M., 2010. A method for function dividing in conceptual design by focusing on linguistic hierarchal relations. J. Comput. Inf. Sci. Eng. 10, 031004.

PART 4

Design Capability and Sociality

CHAPTER 11

Competencies Required for Creative Design Thinking and Their Transfer

Contents

11.1 COMPETENCIES REQUIRED FOR CREATIVE DESIGN THINKING

Chapters 8 and 10 theorized about the basic principles of concept generation and conceptual design. However, these principles are analytical frames that can be used to describe concept generation and conceptual design processes, and are not enough to creatively render these processes. Simply mimicking the arm and leg movements of outstanding athletes does not make one better at a particular sport. In order to improve at a sport, we must acquire *competency* in that sport. In this chapter, the synthetic thinking required to overcome the missing links in the design process will be called **creative design thinking**, the competencies required for which will be discussed. The three competencies are as follows: (1) the competency for determining the boundaries of systems (machine, thinking space, etc.) internally or externally; (2) the competency for abstraction; and (3) the competency for going back and forth in time (Taura and Nagai, 2012). The first two are spatial competencies, while the last one is a temporal competency, all of which will be discussed below.

Creative Design Engineering
http://dx.doi.org/10.1016/B978-0-12-804226-7.00011-9

11.1.1 The Competency for Determining the Boundaries of Systems Internally or Externally

As Chapter 8 elucidates, sequentially more expanding associations while relating concepts to each other leads to the generation of concepts (design ideas) with high originality. Concepts associated during concept generation, regardless of whether they are a direct part of the final design idea, contribute indirectly to design. This section will consider how associations lead to a design idea in creative design thinking from the perspective of how the scopes (boundaries) of associations or thinking are determined. In actuality, how is the scope (boundary) for expanding the association determined? And how is the scope (boundary) for searching for solutions in conceptual design (as discussed in Chapter 9) determined?

Chapters 1–3 discussed the motive of design. It was noted that the motive of design includes inner and outer motives. Regardless of the kind of motive that gives rise to design, there must be some kind of mechanism that determines the scope (boundaries) of design thinking. Generally, boundaries are determined externally from outside the system or determined internally from inside the system. The former corresponds to the requirements or constraints of design mentioned in Chapter 9.

So what does determining the boundaries of design thinking internally mean? Humberto Maturana and Francisco Varela's idea of "autopoiesis" can aid us here. A very simple outline of some relevant information from Humberto and Francisco (1979) is provided below. An "autopoietic machine" is a machine organized (defined as a unity) as a network of processes of production (transformation and destruction) of components that produces the components which (1) through their interactions and transformations, continuously regenerate and realize the network of processes (relations) that produced them; and (2) constitute it (the machine) as a concrete unity in the space, in which they (the components) exist by specifying the topological domain of its realization as such a network. Therefore, autopoietic machines are (1) autonomous, (2) have individuality, (3) specify their own boundaries in the process of self-production, and (4) do not have inputs or outputs. Let us compare the determination of the boundaries of design thinking to those of an autopoietic machine. If this comparison is valid, the following idea emerges. That, in design thinking, the process of sequentially expanding associations while relating concepts to each other is conducted continuously by the process itself, and the concept that will be associated next is determined by the process that precedes this association. This implies that "associations invoke associations," which corresponds to one aspect of creative design thinking. Although the idea of autopoiesis was proposed to determine the essence of living systems, it has since been applied to many phenomena. For example, the creative thinking of artists has been described as autopoietic. This is because the relationship between artists and their works is considered a constant regenerative process.

11.1.2 The Competency for Abstraction

As summarized in Section 10.1, the process of abstraction plays a crucial role in concept generation and conceptual design. Generally, abstraction is the extraction of attributes from a number of concrete concepts. For example, abstracting from the concept of "fire engine" leads to "red" and "something that drives."

Meanwhile, the term "abstract" has another meaning, the one that is implied when we speak of abstract paintings. Let us call this "type 2 abstraction." Type 2 abstraction differs from abstraction as extraction. Abstract paintings are not those that extract attributes from concrete images such as photographs. They are not paintings that simplify whole images. So what are they? Abstract paintings are thought to reflect the desirable form of an object. Chapter 1 stated that "what is desirable" includes that which can be clearly derived from existing problems, as well as the ideal future form that is desirable to pursue. The latter is thought to be related to type 2 abstraction. Chapter 1 noted that the sense of *resonating with our hearts* creates ideality. It can be said that abstract paintings are those, which seek some manner of doing precisely that—resonate with our hearts. This effect of resonating with our hearts cannot be generated by extracting attributes from an existing object in the natural world, but by relying on internal feelings. Creative design thinking particularly requires the competency to perform type 2 abstraction.

11.1.3 The Competency for Going Back and Forth in Time

Chapter 2 stated that the phenomenon of going back and forth in time occurs in missing links. Our selection of base concepts discussed in Chapter 8 also involves this phenomenon. This is because we cannot evaluate the appropriateness of base concepts until we try combining them. Reversal of time is often seen outside design as well. An auxiliary line, such as a median line used in geometry to prove that the two base angles of an isosceles triangle are the same, is an example of this. We are told that the isosceles

Figure 11.1 The problem of finding the reflection point of a beam of light.

triangle theorem can be proven by drawing a median line, but there is no explanation as to where the idea to draw this line came from.

As discussed in Chapter 2, those who can skillfully overcome this phenomenon of time reversal (those who have the competency for going back and forth in time) reduce conflict with social expectations to a minimum, and create designs that contribute to society. The phenomenon of time reversal cannot generally be solved, but can be approached under certain conditions, that is, when certain types of temporal problems are replaced with spatial problems.

Let us consider the problem of finding the reflection point of a beam of light shown in Fig. 11.1.

Regarding paths of beams of light, we know that they take the quickest route to their destination and that the angle of incidence is equal to the angle of reflection when a ray of light strikes a plane surface. If we know only the former, solving the problem would require going back and forth in time. This is because we can know whether a path is the quickest route to the destination only after fixing a temporary reflection point. On the other hand, if we know the latter, we can geometrically identify the reflection point. This discussion suggests that if we can discover the latter knowledge that implies the former one, we can solve the problem of going back in time.

We can thus replace the phenomenon of going back and forth in time with spatial problems. Based on this idea, research has attempted to set a space where design solutions can be searched for with efficiency (Taura, 2008). The design process is sometimes modeled as a solution-searching problem. If modeled thus, there are an unlimited number of possible searching spaces. Within these spaces, creating a space that allows more efficient searching for a superior design solution becomes essential. This research article (Taura, 2008) proposes that the searching space should be set such that similar design solution candidates in the space should receive similar evaluations in the space in which they are being evaluated (Fig. 11.2). Simulation has proved the effectiveness of this idea for neighborhood searching of solutions.

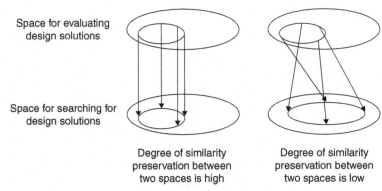

Space for evaluating design solutions

Space for searching for design solutions

Degree of similarity preservation between two spaces is high

Degree of similarity preservation between two spaces is low

Figure 11.2 Similarity preservation between the design solution searching space and design solution evaluation space.

11.2 TRANSFERRING THE COMPETENCIES REQUIRED FOR CREATIVE DESIGN THINKING

11.2.1 Tacit Knowledge

It is difficult to describe the competencies required for creative design thinking using formulae. Such competencies cannot be explicitly described, nor can they be transposed onto a computer. So how can they be transferred? This section will introduce the idea of "tacit knowledge," which may aid toward this (Polanyi, 1966). Michael Polanyi developed his theory about tacit knowledge from the argument "we can know more than we can tell." This argument also applies to the competencies required for creative design thinking. A very simple outline of selected content explained in Polanyi (1966) is provided below. Using the process by which we distinguish faces as an example, Polanyi determines the particulars of the face to be the first term (proximal term) and the whole face to be the second term (distal term), and discusses the relationship between them. He then interprets them from three aspects: the functional aspect, the phenomenal aspect, and the semantic aspect.

- The functional aspect is about how "we rely on our awareness of its features for attending to the characteristic appearance of a face."
- The phenomenal aspect is about how "we are aware of the proximal term of an act of tacit knowing in the appearance of its distal term." "We may say that we are aware of its features in terms of the physiognomy to which we are attending."
- The semantic aspect is about how "a characteristic physiognomy is the meaning of its features, which is, in fact, what we do say when a physiognomy expresses a particular mood."

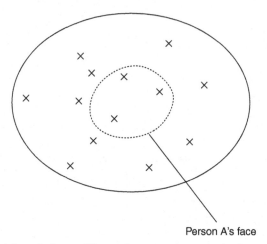

Person A's face

Figure 11.3 Tacit knowledge and first-order concept generation.[3]

Polanyi also discusses the ontological aspect. This aspect may be identified with the comprehensive entity that the proximal term and the distal term jointly constitute. Here, "comprehensive entity" refers to a coherent whole comprised of various particulars (proximal terms), and includes not only so called "entities" such as machines and living things, but also skills made up of a number of linked minute behaviors (such as driving vehicles and playing instruments). Meanwhile, tacit knowledge can be compared to synecdoches (Ito, 1997). This is because the whole face is expressed using the particulars of the face. For example, in set theory, if we take combinations of the particulars of our faces (combinations of the attribute values of particulars such as how the eyes are shaped and the height of the bridge of the nose) as elements, the whole face of a specific person can be expressed by the range of the combinations of attribute values of each particular, or in other words, the subset of elements. Note that the boundaries of such a subset are not explicit. As shown in Fig. 11.3, tacit knowledge can be understood as those cases in which these boundaries cannot be made explicit in the subsets expressing first-order abstract concepts such as those discussed in Chapter 8.

Regarding tacit knowledge, Polanyi (1966) concludes, "scrutinize closely the particulars of a comprehensive entity and their meaning is effaced, our conception of the entity is destroyed," "meticulous detailing may obscure

[3] The dotted line indicates the inexplicit boundary of the subset (although it is specified by the holder of tacit knowledge).

beyond recall a subject like history, literature, or philosophy," and "the belief that, since particulars are more tangible, their knowledge offers a true conception of things is fundamentally mistaken."

So how are the competencies required for creative design thinking acquired and transferred from the perspective of tacit knowledge? Polanyi states, "the destructive analysis of a comprehensive entity can be counteracted in many cases by explicitly stating the relation between its particulars." Additionally, while he states, "where such explicit integration is feasible, it goes far beyond the range of tacit integration," and that, "we possess a practical knowledge of our own body, but the physiologist's theoretical knowledge of it is far more revealing," he also points out, "in general, an explicit integration cannot replace its tacit counterpart. The skill of a driver cannot be replaced by a thorough schooling in the theory of motorcar." He also states, "(I hold the) *active* shaping of experience performed in the pursuit of knowledge… to be the great and indispensable tacit power by which all knowledge is discovered and, once discovered, is held to be true" (emphasis added). Furthermore, regarding the acquisition and transfer of tacit knowledge, Polanyi (1966) uses the terms "interiorize" and "dwell in" to develop his argument thus, "because our body is involved in the perception of objects, it participates thereby in our knowing of all other things outside. Moreover, we keep expanding our body into the world, by assimilating to it sets of particulars which we integrate into reasonable entities… The watcher tries to correlate these moves by seeking to dwell in them from outside. He dwells in these moves by interiorizing them… Chess players enter into a master's spirit by *rehearsing* the games he played, to discover what he had in mind" (emphasis added).

According to the idea of tacit knowledge, in order to acquire and transfer the competencies required for creative design thinking, one should *actively trace* (rehearse) the footsteps of someone who possesses these competencies, or interiorize the competencies through *actively sharing experiences* with such a person.

11.2.2 History Base for Vicarious Experiences of Creative Design Thinking

I would like to introduce an article that is helpful in considering the method for *actively tracing* the footsteps of someone who is outstanding at creative design thinking. This article is about works whose abstract shapes have been created using CAD (Fujihata, 1991). Regarding the relationships between his own works (Fig. 11.4), Masaki Fujihata illustrated the links made by the

Figure 11.4 Example of a product created using CAD (Fujihata, 1991).

commands (operations and inputs given to a computer to generate shapes) used to create these works (Fig. 11.5) and stated the following:

> *I head in directions that I think will lead me somewhere exciting. If I find myself somewhere dull, I either backtrack to my original position or turn a new corner. Arriving at a form, which appeals to me, I stop, or else continue on looking for more new possibilities. This illustration is the printout of the steps I have taken on my journey.*
>
> *I find this map extremely curious. Typically, it is impossible to reiterate a work-process once it has been completed. After a process has flowed, it does not return to its source; hence, the importance of somehow documenting its course. Through photographs and videos, however, you can only reflect on a process, but not repeat it.*
>
> *When I look at this command tree, I rediscover various instances of unexplored possibilities, which have remained open amidst the flow. Or perhaps what I feel is the map chiding me for missing these opportunities.*

Although Fujihata is talking about *actively tracing* his own footsteps, others can also thus utilize this map. In doing so, one can discover what remained hidden in Fujihata's mind, and by interiorizing this, "dwell in" his mind.

Nowadays, the opportunities to design using computers are increasing, and greater possibilities for vicariously experiencing design are emerging.

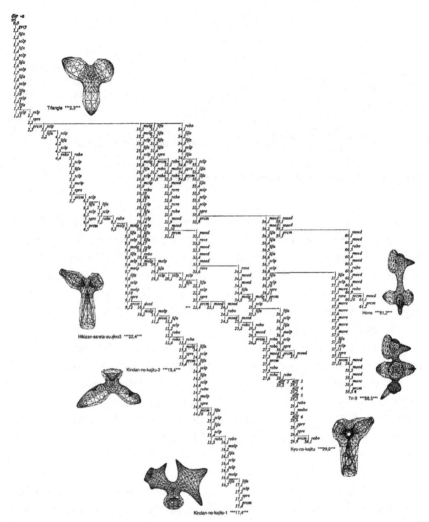

Figure 11.5 A footstep map (Fujihata, 1991).

I would like to introduce the idea of "history base" as one way of doing this (Taura and Kubota, 1999). The purpose of this method is to *actively trace* (vicariously experience) the history of a design while confirming its intention. While the competencies required for creative design thinking are thought to be similar to tacit knowledge, the two differ greatly in certain aspects. While tacit knowledge discusses physical competencies, creative design thinking deals with concept generation and operation. Therefore, when tracing the history of a design, it is crucial not only to reexecute the operation involved,

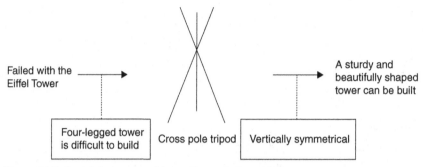

Figure 11.6 Explaining and understanding a design activity.

but also to understand why such an operation was performed, and what the intention of the design was. For example, in chess, when tracing a match between expert players, if we understand why a certain piece was moved, as well as how it was moved in a certain manner, that is, the intention involved, we can more efficiently acquire the skills of such players. So how can we explain and understand the intention for design activities?[1] Let us consider this by referring to Fig. 11.6.

Let us consider the process discussed in Chapter 4, whereby the structure of the spaghetti tower was devised by taking a hint from the cross pole tripod. Assume that we sought an explanation from the person who devised this, such as "Why did you think of making a spaghetti tower like a cross pole tripod?" Let's imagine that he or she answered, "I wanted to create a sturdy and beautifully shaped tower." Although we would be convinced by this response immediately after hearing it, would we not wonder further why the designer chose a cross pole tripod and not some other item? Assume that we then continued to ask, "Why did you choose a cross pole tripod to create a sturdy and beautifully shaped tower?" Imagine that his or her response was, "Because I thought a vertically symmetrical shape would be a good idea." We can thus conclude that the designer "thought of making the spaghetti tower like a cross pole tripod, because they thought that this was sturdy and beautiful, because this was a vertically symmetrical shape." This kind of explanation is known as a teleological explanation.

On the other hand, let us assume that in response to our question, "Why did you think to make a spaghetti tower like a cross pole tripod?"

[1] In this book, the mental operation of concepts (eg, concept generation) and external operation of products (eg, creation of shape) by designers are collectively referred to as design activities.

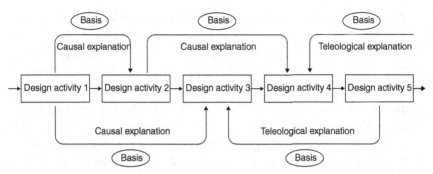

Figure 11.7 Explaining design activity and describing motive based on links and their attributes.

the designer had originally responded, "Because I built a spaghetti tower like the Eiffel Tower in the past and failed." Although we would be convinced by this response immediately after hearing it, would we not wonder further whether having failed at creating a spaghetti tower like the Eiffel Tower necessarily implied that the designer needed to choose a cross pole tripod? Let us say that we then continued to ask, "Why did failing at building a tower like the Eiffel Tower prompt you to choose a cross pole tripod?" Imagine that his or her response was, "Because it was difficult to build a four-legged structure." We can conclude that the designer "focused on the three-legged structure of a cross pole tripod because the previous attempt to create a spaghetti tower like the Eiffel Tower failed, because a four-legged structure like the Eiffel Tower was difficult to build." This kind of explanation is known as a causal explanation.[2]

The process involved in such explanations can be summarized as follows. With regard to the design activity of focusing on a cross pole tripod, the first question asked is "why?" followed by the second "why why?" The responses to the first "why?" in both cases, namely, "I wanted to build a sturdy and beautifully structured tower," and "I previously failed at building a spaghetti tower like the Eiffel Tower," list the explanations for the design activity of "focusing on a cross pole tripod." The responses to "why why?" in both the cases, namely, "I thought a vertically symmetrical shape would be good," and "building a four-legged structure was difficult," are those that lie behind the explanations for the design activity of the person, and can be considered the feelings and criteria, or awareness of problems that led to the motive of design.

[2] Teleological explanations and causal explanations are explained in detail by von Wright (1971). This section discusses the scheme for describing explanations and motive of design activities based on von Wright's arguments.

From this discussion, we can consider that, as regards a particular design activity (A), by pointing to the activity within the activities that were performed previously that explains the cause for activity (A), and the basis that explains that cause, along with the activity within the activities performed thereafter that explains the objective of activity (A), and the basis that explains that objective, we can articulate the intention for the activity. In other words, as shown in Fig. 11.7, if we explain an activity using its links with other activities and also describe the basis that explains those links, we can trace the activity while understanding the intention behind it.

The Activity Chain Model is proposed on the basis of the aforementioned discussion (Taura et al., 1999). In this model, the history and intention (explanation and its basis) of a design are described in terms of chains between design activities. The Activity Chain Model makes it easy to grasp the history and intention of a design and enables the building of a work environment for design with a high explanatory potential that is involved in the idea of the history base. In the future, this is expected to become an effective method for supporting cooperation among design teams dispersed worldwide, and accumulating and passing on experiential knowledge that remains hidden in the field of design.

Task

In the future, do you think computers will be able to design? If so, what kind of method do you think they will use? Consider the following two cases.

1. Mimicking us
2. Designing using a method different from us

REFERENCES

Fujihata, M., 1991. Forbidden Fruits. Libro Port Co., Ltd, Tokyo.

Humberto, R.M., Francisco, J.V., 1979. Autopoiesis and Cognition: The Realization of the Living. Kluwer, Holland.

Ito, M., 1997. Tacit knowledge and knowledge emergence. In: Taura, T., Koyama, T., Ito, M., Yoshikawa, H. (Eds.), The Nature of Technological Knowledge. Tokyo University Press, Tokyo, (in Japanese).

Polanyi, M., 1966. The Tacit Dimension. Peter Smith, Gloucester.

Taura, T., 2008. A solution to the back and forth problem in the design space forming process: a method to convert time issue to space issue. Artifact 2 (1), 27–35.

Taura, T., Kubota, A., 1999. A study on engineering history base. Res. Eng. Des. 11 (1), 45–54.

Taura, T., Nagai, Y., 2012. Concept Generation for Design Creativity: A Systematized Theory and Methodology. Springer, London, pp. 21–26.

Taura, T., Aoki, Y., Takada, H., Kawashima, K., Komada, S., Ikeda, H., Numata, J., 1999. A proposal of the activity chain model and its application to global design. Adv. Conc. Eng., 29–38, Bath.

von Wright, G.H., 1971. Explanation and Understanding. Cornel University Press, New York.

CHAPTER 12

From Product Design to Technology Design

Contents

12.1 THE RELATIONSHIP BETWEEN SCIENCE AND TECHNOLOGY AND DESIGN

As discussed in Chapter 1, design can be said to act as a bridge between science and technology,[1] and society, through the creation of products. Building this bridge requires various types of ingenuities. First, the ingenuity for incorporating science and technology into products is required. The behavior of physical phenomena cannot be fully grasped by simply observing them in the laboratory. Unexpected behaviors not observed in the laboratory may arise when products with science and technology incorporated are situated in the real world (this will be discussed in detail in Chapter 13). Therefore, the behavior of products should be controlled to comply with our intentions in the fields where they are actually used.

The term "technology" is sometimes used to refer to the knowledge and skills required to apply the basic principles of physical phenomena to the real world. In this book, such knowledge and skills will be treated as "technology in the narrow sense," and when the meaning is broadened to include the incorporation of science and technology into products, this will simply be called "technology." Science and technology incorporated into products then interacts with users in society. Efficient, smooth, and

[1] In this book, as discussed in Chapter 1, the term "science and technology" is used as a generic term for an understanding of the basic principles of physical phenomena and the basic knowledge required to apply it.

Creative Design Engineering
http://dx.doi.org/10.1016/B978-0-12-804226-7.00012-0
145

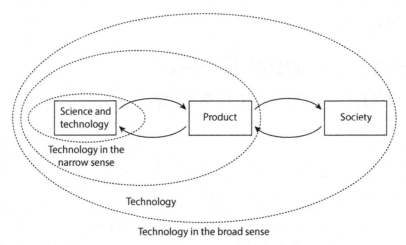

Figure 12.1 The scope of the meanings of "technology" in design.

attractive interactions are desirable, but this also requires ingenuity. Therefore, when taking into account the manner in which products interact with society, the term "technology in the broad sense" is used.

This relationship between science and technology, products, and society, is illustrated in Fig. 12.1.

Recently, there have been amazing developments in science and technology. In this chapter, the kind of science and technology that has rapidly developed over the past few decades and whose resultant products are being increasingly used in society will be called "highly advanced science and technology." On the other hand, science and technology that has already been systematized will be called "conventional science and technology." As examples of the former, this chapter will look at genetically modified foods using genetic modification science and technology and nuclear power stations based on nuclear power science and technology, while as examples of the latter, it will look at the steam locomotive and motorcar[2] using classical mechanics.

There is a difference between designs based on conventional science and technology and those based on highly advanced science and technology, in terms of the relationship between the locations where the products themselves are developed and those where the science and technology incorporated in them is researched. In designs based on conventional science

---

[2] Since recent cars use highly advanced science and technology for energy saving and safety enhancement, they could also be categorized under highly advanced science and technology. However, here we are focusing on basic motor vehicle technology and have categorized them as conventional science and technology.

and technology, the location of developing the product itself and that of researching the science and technology used in the design are closely related to each other. For example, in the case of steam engines, the person who was researching the principles of the steam engine developed the first one (James Watt). In the case of cars, although the research and development work has been divided, both are often conducted by the same company. Then science and technology and the products that incorporate the findings of the research make an impression on society (cause anxiety and expectations), and are evaluated by society as a combined object. This is exemplified by the fact that the principles of steam engine and steam locomotive are seen as almost one and the same.

In contrast, in the case of designs that are based on highly advanced science and technology, the organizations that research the principles underlying these designs and the organizations that develop products based on the knowledge of such principles are often different. Then, the objective of these designs is to implement this knowledge of principles into the products. For this reason, highly advanced science and technology receives direct feedback from society about the science and technology itself. For example, society may exhibit anxiety over genetic modification or nuclear power science and technology.

As discussed, structures of designs using conventional science and technology differ from those using highly advanced science and technology. Figs. 12.2 and 12.3 illustrate the relationship between science and technology, products, and society for each type of science and technology.

12.2 INTERACTIONS OF SCIENCE AND TECHNOLOGY AND ITS PRODUCTS WITH SOCIETY

Products that incorporate science and technology interact with society in the post-design phase. Products provide society with services such as utility or a sense of satisfaction, but the result does not always necessarily add value to society. They can also cause accidents or damage. As discussed in Chapter 2, it seems that in many cases, products are used by society, accidents occur, society reflects upon these, and somewhere within the repetition of this cycle, the product is (or appears to be) accepted by society. In this section, we will call the situation in which a product is (or appears to be) received for use by society as *accepted* by society. Then, as a result of the product being accepted by society, the implicit or explicit feelings and criteria, or awareness of problems formed by society lead to the social motive of design for the next product.

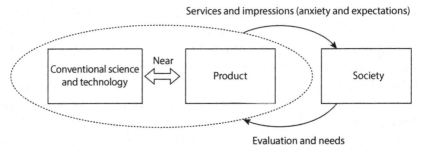

Figure 12.2 Structure of design based on conventional science and technology .[3]

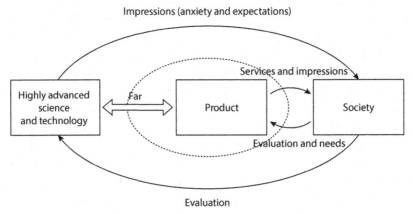

Figure 12.3 Structure of design based on highly advanced science and technology .[4]

So what does it mean to say that science and technology and the products that incorporate it are accepted by society? In order to discuss this, we must not only look at superficial phenomena such as whether the product is actually being used, but also turn our attention to the interactions between the product and society at a deeper level. This can be considered from the following main viewpoints: (1) whether specific science and technology in use has permeated society as *art and culture*, (2) whether *social systems* have been built to respond to the downfalls of specific science and technology in use, and (3) how the principles of specific science and technology in use (in a broad sense) have been *cognitively perceived* (Taura, 2014).

---

[3] The dotted line represents the main scope of design.
[4] The dotted line represents the main scope of design.

12.2.1 Science and Technology, and Art and Culture

Science and technology, and its products, are sometimes interpreted by society as art and culture. Actually, science and technology, and its products, are often depicted by artists. For example, the famous Impressionist painter Claude Monet painted the steam locomotive. Let us focus on the fact that this was painted by a famous Impressionist painter. The painting is indicative of the steam locomotive being accepted by society on a deep emotional level.

The Japanese sword is a similar example. The Japanese sword is made using specialized technology, including heating, hammer forging, and hardening. Today Japanese swords are traded as works of art, for their clear, pure, cold beauty.

Incidentally, examples of the steam locomotive and the aforementioned Japanese sword belong to the category of conventional science and technology, and the working principles and manufacturing processes can been understood visually. Furthermore, when we see a steam locomotive, our hearts dance at the dynamic movement of the steam engine, and when we see a Japanese sword, our hearts are moved by its beauty. This illustrates how science and technology can evoke new human feelings. It also indicates the close relationship of science and technology to art and culture, and furthermore, its potential to act as art and culture itself for society. The penetration of science and technology into society as a form of art and culture can be considered as one aspect of acceptance of science and technology and its products by society at a deep level.

However, when it comes to highly advanced science and technology, a majority of the products continue to remain distant from art. This is because much highly advanced science and technology exceeds the scope of our natural senses. However, for instance, even if science and technology itself is not depicted in paintings, the products in which it has been incorporated sometimes permeate society as culture. For example, although we rarely see the working principles and manufacturing processes of televisions and computers with our own eyes, a huge number of images are transmitted and new cultures formed through these technologies. Furthermore, recent cellular phones allow users to manipulate the screen by dragging and expanding or collapsing the screen between their two fingers. These manipulations come *naturally* to us, but they are not performed in our *natural world*. It seems that the power of highly advanced science and technology has broadened the scope of "feeling-based" or "analog" operations.

12.2.2 Science and Technology, and Social Systems

Products in which science and technology have been incorporated often cause losses to society. For example, cars continue to cause many traffic accidents, even now. Why might society continue to use cars when they pose such a high level of danger? This is likely because the positive effects of cars are large enough to allow us to accept a certain amount of loss. However, for instance, if no measures had been taken against such losses, society may not have accepted the product. Social systems are often established for science and technology and its products to deal with losses they cause. For example, in the case of cars, the responsibility[5] for traffic accidents is divided between manufacturers, users, and road managers, and this is further divided into the responsibility held by individuals and companies. Measures to manage such losses are termed as "risk management"; the idea of risk management could also be described as having been created to conveniently manage the psychological anxieties of fear and danger in a quantitative manner, which were not quantifiable by themselves.

However, in case of highly advanced technologies, existing social systems are unable to deal with the very high dangers they pose. For example, genetically modified foods and nuclear radiation might cause damage to future generations, our children, and grandchildren. Conventional notion of responsibility seems to have been formed among people of the same generation. So what does it mean to have a responsibility toward our children and grandchildren? If damage does occur, the person responsible would no longer be around to take responsibility for it. In order to deal with this situation, a need arises for new notions that expand our conventional way of thinking about responsibility.

Meanwhile, it can be said that social systems required to deal with shortcomings of products are established through accidents. The process of reflecting upon accidents makes the danger and scope of responsibility clear. However, a number of products incorporating highly advanced science and technology that might pose extremely high danger, such as nuclear power, cannot be permitted to cause accidents, as the damage would be immense. If accidents are not allowed to occur, this means that danger and scope of responsibility are determined at a desk, without actually experiencing them. This is extremely difficult. Note that this problem also arises with conventional science and technology, if the product is produced in too

[5] According to Tomonobu Imamichi, a researcher in aesthetics and philosophy, the term "responsibility" has only been in use since the Industrial Revolution.

large quantities. For example, phenomena such as climate change, which are caused by the release of carbon dioxide, have the potential to inflict immense and irreparable loss to future generations.

12.2.3 Science and Technology, and the Cognition Process

How are science and technology, and the behavior of their products cognitively perceived by us? Many aspects of conventional science and technology are perceivable to us. For example, we can perceive the movement of a steam locomotive. Although we cannot see directly into the steam boiler, we can experience the working of its principles by looking at the turning wheels and smoke emission. This is similar for car engines. Through our experiences, we perceive science and technology and its products in our minds, which make good impressions in some cases and bad impressions in others.

However, most highly advanced science and technology is not directly perceivable. We cannot perceive radiation. Nor can we see DNA with our naked eyes. How can we cognitively perceive such physical principles that exceed the limits of our natural senses?

Perhaps one way we cognitively perceive these is *indirectly*, as a *notion*. Even if the principles of a physical phenomenon are difficult for us to perceive, we can understand the physical phenomenon as a notion. When designing the structure of a product (a machine), we calculate "forces," but force itself in classical mechanics cannot be observed. What is observed is "displacement," "position," "velocity," and "acceleration," and "force" is derived from these using the appropriate laws. "Force" can be thought of as a hypothetical notion created by us. We feel as though we understand this notion, and we can also use it to feel as though we understand physical phenomena. In other words, by cognitively perceiving (in the broad sense) the notion of "force," we can say we have understood the working principle of a product. On the other hand, the debate over radioactive waste involves discussing how it will be stored for hundreds of thousands of years. How should we interpret this time scale? When the time scale of the future is analogized with that of the past, then we feel like we have understood "future." As discussed in Chapter 1, "future" is an extremely abstract notion that can only be expressed through language. While we can picture the future form of say "future Tokyo" or "our future lifestyle" at some point in the future, we cannot picture the future itself. Only when the time scale is analogized with that of the past, we will feel as though we have understood the future.

To summarize, we are able to feel like we have understood a notion that is quite distant from the real world, using which we can (feel as though we can) cognitively perceive (in the broad sense) phenomena that are difficult to perceive. In the case of highly advanced science and technology, it seems that we need a new system of notions in order to understand such difficult-to-perceive phenomena.

Meanwhile, even when their principles have not been sublimed into notions, difficult-to-perceive physical phenomena are sometimes (thought to be) accepted experientially. Some examples are microwaves and induction cookers. Although various dangers have been identified, on the whole, we can say that these have been accepted by society. While it is likely that scientific explanations regarding their safety have reassured us, the experience of repeated use could also be said to have played an important role. Perhaps somewhere within the process of repeated use without any particular accident or injury, we ceased to be aware of the danger involved.

12.3 ANALYTICAL AND SYNTHETIC ACCEPTANCE

As discussed in the above section, for science and technology and its products to be accepted by society, the use of products must be *experienced* by users. Within this experiential process, society accepts the utility of the science and technology and the product, and then internalizes and interprets these as art and culture. Meanwhile, this experience may include accidents. Through the experience of accidents, social systems are established to deal with its shortcomings and consequently, trust toward the product increases.

Such acceptance of science and technology and its products by society based on interactions with those products will be called **analytical acceptance** in this book. When the interaction with a product is analyzed, it enables sharing of the feelings and criteria, or awareness of problems relating to the product. For example, an analysis of the occurrence of an accident gives rise to new ways of understanding the product, which in turn may result in the establishment of new insurance systems.

Meanwhile, highly advanced science and technology have the potential to cause huge, irreparable damage, and thus cannot be permitted to cause accidents. In such cases, the products are accepted by society without their danger being experienced firsthand. Furthermore, for products yet to exist, society is forced to determine whether it should accept these products at a desk. Examples include science and technology on radioactive waste treatment plants and genetically modified foods.

Such acceptance of products into society at a stage when the product has yet to come into existence or there are yet to be any experiences of its use will be called **synthetic acceptance** in this book. In synthetic acceptance, the utility and dangers associated with the product must be assumed. Moreover, the feelings and criteria or awareness of problems are required to be shared within society based on such assumptions. This is extremely difficult. This is because (as will be explained in Chapter 13) it is not possible to predict everything about the behavior of a product. In other words, *unanticipated* events are unavoidable. Furthermore, even if most aspects about the behavior of a product could be predicted, human beings find it difficult to feel like we understand something without having experienced it firsthand.

However, even without firsthand experience, there is some possibility that science and technology and its products will be understood using the following method. First, as discussed in Chapter 3, inferring latent functions and latent fields enables the systematic assumption about the utility or danger of products. Next, the method involves the application of virtual reality technology, which has been making rapid advances in recent years, in order to virtually experience the assumed behavior of products. Finally, as discussed in Section 12.2.3, we can devise new methods by taking a hint from the idea that we can understand objects or things as notions without directly experiencing them with our senses.

12.4 DESIGN OF TECHNOLOGY

The discussion in the previous section can lead us to the idea of **design of technology** (Fig. 12.4). Traditionally, design was primarily concerned with the quality and cost of products (function, structure, shape, etc.) and their manner of creation. In contrast, design of technology broadens this perspective to include the relationship between science and technology, and society. Here, products become tools for design of technology. The *role of design* then becomes "to bridge the gap between science and technology and society at a deep level, through the creation of products," and the discussion includes the permeation of science and technology and its products into society as art and culture, the indirect cognition of physical phenomenon and time scales that cannot be directly experienced by our natural senses, and the establishment of social systems to deal with the shortcomings of products.

Based on the definition of design provided in this book, design of technology can be identified as the process of composing the desirable relationship between science and technology and society in a multifaceted manner

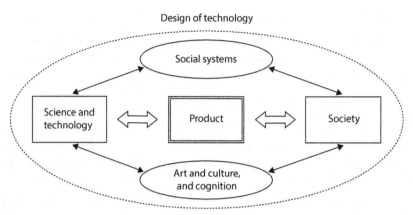

Figure 12.4 Design of technology.

toward the future, taking into account art and culture, and social systems. This is perhaps no different from the definition of "engineering" itself. Traditionally, main activities bridging science and technology with society have consisted of simple explanations, using charts and diagrams, of principles and functions by those who are experts in a particular science and technology and/or its products, or the use of statistical methods in explanations.

However, as discussed in this section, for science and technology and its products to be accepted into society at a deep level, or for their rejection to be determined, those who are engaged in researching or developing science and technology and its products must interact with experts from a large number of fields, such as artists, novelists, sociologists, and philosophers. Given this, what role might the designer play? The designer himself/herself need not become an artist, sociologist, or philosopher. The designer must instead prepare himself/herself mentally. This will be discussed in the next and final chapter.

Task

Choose a product from the following list, investigate its scientific/technological and cultural/artistic characteristics, and discuss the relationship between them.

Five-storied pagoda, Dancing Satyr, and Notre Dame Cathedral

REFERENCE

Taura, T., 2014. The design of technology: bridging highly advanced science & technology with society through the creation of products. In: Taura, T. (Ed.), Principia Designae: Pre-Design, Design, and Post-Design: Social Motive for the Highly Advanced Technological Society. Springer, Japan.

CHAPTER 13

The Meaning of Creative Design Engineering

Contents

13.1 WHY DO WE DESIGN?

This book has discussed the question, Why do we design? focusing on the motive of design. The creation of innovative products is a mandate for the present and the foreseeable future. Meanwhile, highly advanced science and technology and the products that incorporate it have the potential to bring about accidents that cause irreparable damage. Thus, our discussion needs to go back to the origin of products.

Motive of design was defined as "the reason behind the design process, arising from feelings and criteria, or awareness of problems relating to a product," and Chapters 2 and 3 considered the motive of designers, while Chapter 12 discussed the motive created in the process of accepting products and the science and technology underlying them.

Nevertheless, it seems that the motive of design of many products is unclear. Many products in the world exist without an explicit reason. Cars, as discussed in Chapter 2, are an example.

The inability to identify the motive of design does not emerge from neglect, but rather results from the difficult to identify nature of the motive of design. For example, the question, Why do we talk? is meaningless. Sometimes we talk with some kind of intention, and at other times, we talk with no particular intention. Similarly for design, we engage in design as an unconscious act. Furthermore, the discussion in Chapter 3 suggests that one reason for the constant unconscious creation of products may be that

Creative Design Engineering
http://dx.doi.org/10.1016/B978-0-12-804226-7.00013-2

man-made objects resonate with our hearts and our hearts also dance at the act of creating such objects, both of which are thought to be phenomena without any particular intention.

However, design differs critically from the act of daily conversation in terms of its tremendous effect on society. As will be discussed below, there are no absolutes in product safety. The safety level of products may increase with improvements; however, ensuring the safety of a product to perfection is not realistic. Therefore, when such products are designed and used, we should *prepare ourselves mentally* for the possibility of huge accidents. Such preparation requires that we clarify the social motive of design. However, the following paradoxical problem now faces us. Because the danger that highly advanced science and technology and its products might carry is extremely massive, they cannot be permitted to cause accidents. This means we may not have the opportunity to experience firsthand the possible dangers of these products. Can we prepare for danger that we have never experienced? It might be difficult for society to be aware of the problem of worst-case scenario accidents without having experienced them. Note that this argument is not limited to products of highly advanced science and technology, but also applies to those of conventional science and technology. One example is the problem of climate change, which was discussed in Chapter 12.

Furthermore, for products that have not yet come into being, society must sometimes determine whether these products should be accepted without actually experiencing them. Examples include radioactive waste treatment plants and genetically modified foods. In such cases, we must *assume* the utility and danger of products. Moreover, the feelings and criteria or awareness of problems must be shared within society based on such assumptions, which is extremely difficult. This is because, as will be discussed below, it is not possible to predict everything about the behavior of a product.

In modern times, where science and technology have made great strides, the need to carefully consider the motive of designing products has increased in importance. However, as discussed, the motive of design is fundamentally difficult to identify. Given these conditions, in cases where motive cannot be clarified despite going to the greatest lengths to do so, particularly where such a product cannot be permitted to cause accidents, does this imply that the product should not be made?

Note that this does not mean that it is okay for products that have already been made to cause accidents. Needless to say, efforts should always be made to increase the level of safety of products.

13.2 WHY ARE WE ABLE TO DESIGN?

This book has discussed the question, Why are we able to design? by focusing on the missing links in the design cycle. Missing links in the pre-design phase are noncontinuities to be overcome when the social motive of design that would lead to the creation of a new product is not fully developed. Missing links in the design phase are noncontinuities to be overcome when no similar products exist, and the requirements or specifications for generating the new product are only expressed in abstract terms. Finally, missing links in the post-design phase are the acceptance of products into society before such products exist or before accidents caused by such products are experienced. It was also noted (Section 2.4) that a time reversal occurs within these missing links, and synthetic methods for overcoming missing links were discussed.

With regard to the pre-design phase, synthetic methods of concept generation were discussed. Specifically, Chapter 3 showed how inferring latent functions and latent fields could serve as an effective tool for overcoming missing links. Chapter 8 discussed high-order concept generation, which involves the sequential association of concepts focusing on dissimilarity, based on inner motive.

With regard to the design phase, Chapter 10 discussed synthetic function decomposition through the mixing of fragmental knowledge.

With regard to the post-design phase, Chapter 12 discussed synthetic acceptance through the following: inferring latent functions and latent fields, understanding of objects or things as notions, and virtual experience, even when products cannot be directly experienced by the senses. Furthermore, regarding the discovery of uses for products outside those intended by their designers, it was shown in Chapter 3 that inferring latent functions and latent fields is an effective method.

What is common to synthetic methods is their focus on our processes as human beings to manipulate *concepts or notions*. Meanwhile, three competencies that are required to deploy these processes of manipulation of concepts or notions were discussed: (1) the competency for determining the boundaries of systems internally or externally, (2) the competency for abstraction, and (3) the competency for going back and forth in time.

To summarize the aforementioned discussion, the reason we are able to design innovative products that are in agreement with the needs of society is that we effectively overcome the missing links in the three phases of design based on three competencies.[1] The sum of these three competencies

---

[1] The missing links in the post-design phase cannot be directly overcome by the designer, but they are included here as they play a crucial role in the interactions between people and products.

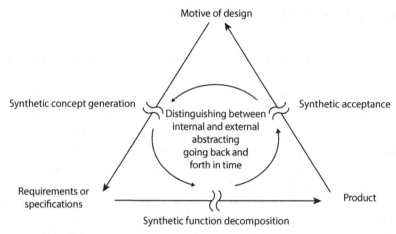

Figure 13.1 A schematic representation of design capability.

and methods for overcoming the missing links in the three design phases could in fact be called the *design capability* required to creatively deploy the non-continuous design cycle model. Fig. 13.1 presents a schematic diagram.

Note that analytical methods also exist along with synthetic methods as discussed in Chapter 1. It is important to emphasize that these are also essential and play a vital role in design.

13.3 THE MEANING OF DESIGN IN ENGINEERING

As discussed in Section 12.4, design of technology requires interactions with experts from a large number of fields, such as artists, novelists, sociologists, and philosophers. So what role do designers play in such interactions? The designer need not become an artist, sociologist, or philosopher. What is important is the mental preparation of the designer. Specifically, an attitude that emphasizes *meaning* instead of the *procedure* of design is required.

13.3.1 What is Correctness in Engineering Education?

In engineering education terms, what does it mean to be *correct* or *incorrect*? First, let us consider correctness and incorrectness, as we understood these terms in science and mathematics at school. "I have a sister who is 2 years old. I am 4 times the age of my sister. How old am I?" Of course, the correct answer is 8 years old. However, let us assume that one student answered as follows, "My sister is 3 years old. Therefore, there is something wrong with this problem. If I multiply my sister's age by 4, the answer is 12, but I am now 10 years old. This problem does not make sense." The teacher would probably respond as follows, "This problem is *presuppositional*. Assume that your sister is 2 years old." However, the student may become more and more confused. "What does *presuppositional* mean?"

In actuality, most of what we learn in mathematics and science lessons in school takes place within a *presupposed world*. Some may think this is valid for mathematics, but not for subjects such as physics and chemistry, which involve experimentation. However, even in experiments, the term "presupposed" would apply to most cases. This is not because the problem takes a presupposed form, but because the ideas themselves are presupposed. As discussed in Section 12.2.3, "force" as used in classical mechanics is a hypothetical notion. This is valid for "particles" as well. We learn that particles have mass but no volume. The knowledge system of mechanics used in engineering is based on this idea of particles. However, what does it actually mean to have an entity with mass but no volume in reality? This is difficult to imagine based on our everyday senses. In other words, particles exist in a "let us presuppose that…" hypothetical world. As such, engineering is built upon the idea of "let us presuppose thus—this will allow us to explain the phenomenon."

However, we have the odd habit of considering a hypothetical statement to be true if it provides a good explanation for reality. Think, for example, of Hooke's law. This law measures the relationship between the amount of elongation of a spring and the weight hung on it, and confirms that this relationship is linear. However, let us assume that when tested in an experiment, the relationship was not linear. We may thus think, "this experiment was wrong." This is because we are led by our preconception that it should be linear, and thus end up rejecting the results of our experiment. However, in fact, there was nothing wrong with the experiment itself. If we limit our discussion to this point, we should be able to say that our experiment was correct. Furthermore, even if our experiment had indicated a linear relationship, a strictly linear relationship would be impossible. We account for the deviation from the line as "error," which refers to the difference resulting from our mistakes. This is because we believe that no deviation from the line is correct, and deviation from the line is incorrect. However, considering that such deviation did occur in actuality, there is no doubt that a deviation from the line is correct. Let us refer to another example. A gas with no volume and no intermolecular force is called an "ideal gas." What does "ideal" mean here? It likely refers to the simplest possible description of the gas. This can then imply that we consider the simplest possible description of a state as "what is desirable."

As discussed, good explanations for reality are often confused and interchanged with truth (correctness). Heliocentric theory certainly provides a better explanation (can explain more phenomenon more simply) than does Ptolemaic theory. However, this does not necessarily mean that we cannot reject our perception that the sun appears to revolve around the earth when we see sunrise and sunset with our naked eyes.

In engineering education, what are we normally taught as correct? We are taught the correctness of *procedures* for the presupposed world that can explain reality well. We are trained in the correct procedures for solving equations and accurately calculating strength, and if there are no errors in that procedure, the problem is judged to have been solved correctly.

Note that the presupposed world for understanding physical phenomenon can also be described as the prerequisites for determining these procedures. The content of these procedures depends on how we define the presupposed world.

13.3.2 The Limitations of Predicting Product Behavior

Since products to be designed do not exist in the world yet, we must predict how they will behave before they are used in society. However, it is not possible to do this with perfect accuracy. In other words, we will never be able to perfectly ensure the safety of products. This will be explained below.

In order to predict product behavior, we must presuppose the field in which the product will be used. Then, the behavior of the product can be predicted for that presupposed field.

First, let us think about coming up with presuppositions about the field. As discussed in Chapter 3, fields include physical fields, contextual fields, and semantic fields. We can fix any temporary field we like with a product. However, when the product is actually used in society, the field of its usage may not fall within the scope of the fields that have already been previously temporarily fixed. For example, when designing an airplane, the physical state of the aerospace, such as the airflow and temperature, are presupposed. However, we cannot determine with certainty whether the physical conditions in which an airplane will fly will fall within our presupposed scope until it has actually been flown a number of times. We base our presuppositions on existing information. However, we should know that previous conditions will not necessarily replicate themselves. Furthermore, in the case of contextual fields and semantic fields, these are determined based on our own behaviors, making the future difficult to predict. This is because, as discussed in Chapter 3, users may use products in fields and impart meanings to them that had not been originally intended.

Now, even if we were able to presuppose the fields of usage of products, it is impossible to accurately predict the behavior of products. This is because the knowledge we have gained from physical phenomena is nothing but *knowledge used to explain* such phenomena. The behavior expected of products is often something that does not yet exist in the natural world itself. Thus, in order to create new behaviors, we must *reconstruct* the knowledge we have so far gained. However, such reconstruction results in *gaps* between actual and predicted phenomena, the reason for which is explained below.

First, the concept structure required to provide explanations and the concept structure required for design are fundamentally different (Section 5.2.2). Nothing new is born of the knowledge that seeks to explain. If we were to forcibly attempt to think of new things from the knowledge that provides explanations, gaps would arise between actual and predicted phenomena.

Second, in order to understand physical phenomena, we sometimes articulate their principles as hypothetical notions (Section 12.2.3). If we attempt to predict new behaviors from our articulation of hypothetical notions, gaps would arise between actual and predicted phenomena.

Third, because the knowledge that seeks to provide explanations is created based on limited perspectives, behaviors outside the scope of such knowledge cannot be reliably predicted. For example, linear approximation provides a range of applicability. Of course, we cannot make predictions outside this range. Moreover, when multiple physical principles operate, interactions may arise between them. However, interactions cannot be described with merely the knowledge of each of these areas. For example, behavior that arises through interaction between mechanical and chemical behavior cannot be predicted merely using the knowledge that seeks to explain either mechanics or chemistry alone. Limiting perspectives that emerge from the classification of disciplines is a common characteristic that can be found across academia, and Yoshikawa comments on this as follows (Yoshikawa, 1993), "Dividing academic domains makes ensuring compatibility within theories which belong to each domain easier, and subdomains are created within the same domain due to simplification of theories." He also states, "However, what is even more important here relates to the relationships between domains. That is compatibility is not sought between theories that belong to different domains. In any case perspectives are things that have been historically created based on the idea that phenomena that occur between differing perspectives are independent of each other at least as functions that have meaning to us. Therefore it is only rational that, if domains are created based on perspectives, phenomena that belong to different domains are independent of each other."

Based on this, it is impossible to accurately predict the behavior of a product in the field in which it is placed. Therefore, full-scale or reduced-scale prototype models are normally created, the behaviors of which are investigated, and improvements are made to the product until it behaves as intended. Furthermore, prototype models are used in the actual field to test whether the field presupposition is acceptable or whether problems arise in terms of the behavior predictions. Even so, when the product is actually used, unanticipated events can occur. In some cases, accidents may occur. When this happens, a thorough investigation of the cause is conducted, the field presupposition is reevaluated, and improvements are made to the product structure. If this cycle is repeated appropriately, the product behavior can be almost completely controlled provided it is used in that particular field.

13.3.3 Traits Required of Those Who Design

Engineering usually only takes responsibility for the *procedures* of design. If an accident occurs, the engineer is only held responsible for errors in the procedures, such as miscalculations about the strength. If errors are found with the procedure, the design is termed incorrect. On the other hand, if no errors are found in the procedures, then the design is deemed correct or problem free. The mandatory prerequisites for these calculations are generally provided from outside, and whether or not they are correct is outside the scope of the engineer's responsibility. Incidentally, the same idea can be observed for formal logic as well. In formal logic, something implied from a false proposition is taken to be true. In other words, regardless of whether the proposition is true or false, the act itself of implying a proposition from a false one is thought to be true.

So who decides the prerequisites for design, and how? Let us consider this based on the design cycle model discussed in this book. In the pre-design phase, motive of design becomes the prerequisite for design; in the design phase, the physical phenomena related to the product behavior and the presupposed world used to understand those physical phenomena, as well as the presupposed constraints used when manufacturing the product, become the prerequisites for design; and in the post-design phase, the field of product use becomes the prerequisite for design. For example, when designing a bridge, the motive for needing that bridge becomes the primary prerequisite, and the physical phenomena arising from the bridge such as fatigue and erosion, the theory used to calculate its strength, and the design criteria, etc. become the secondary prerequisites. Furthermore, the tertiary prerequisites include the magnitude of earthquakes predicted for the area surrounding the bridge (physical field), the predicted traffic levels (contextual field), and the symbolism created by the bridge (semantic field). Here it must be noted that in addition to explicit prerequisites, many implicit prerequisites exist within society.

The prerequisites for design appear as the *meaning* of design. In terms of motive of design, the need for the product is equivalent to the meaning of design. In terms of the presupposed world for understanding physical phenomena and the manufacturing methods of the product, the models of physical phenomena at work in the product act as the meaning of design, and in terms of the field in which the product is used, the utility and danger for society is equivalent to the meaning of design. Here, the first and third can be called the **social meaning of design**, and the second, **physical meaning of design**. Let us assume that a designed bridge that was used by many people was hit by an earthquake one fine day and the cars on it were plunged into the river. In this case, in addition to the initial meaning—a product for crossing the river—the bridge now connotes susceptibility to earthquakes, and while convenient, is also dangerous.

Until now, engineering has utilized physical phenomena to create various products. Thus, it has been oriented toward accurately understanding physical phenomena and skillfully putting these to use. The majority of such knowledge of physical phenomena is described in procedures, which are then simulated using computers. This is valid for the structure and shape of cars and the earthquake proofing of buildings. However, we are the ones who make and use products. The question of how a product should be is ultimately debated in terms of the social meaning of design. Engineering should not adopt an indifferent stance to what it does and the meaning of the products that are created as a result. We must be humble about design. In particular, we must bear in mind that we cannot predict everything about the behavior of products and the fields in which they will be placed. Unanticipated positive effects may arise, but unanticipated accidents may also occur. We must know that it is impossible to describe all the design prerequisites, and *unanticipated events will always occur*. While the procedures of design should be followed accurately, we should not focus only on procedures, but also consider the meaning of unanticipated behaviors and fields, attempt to carefully prevent accidents, and respond quickly if such accidents do occur.

> The designer must not restrict his/her perspective to the procedures that they themselves carry out, but think about the meaning of design, and communicate this to other stakeholders while also listening to what these stakeholders say.

In particular, designers should interact directly or indirectly with experts from a large number of fields, such as artists, novelists, sociologists, and philosophers.

Meanwhile, procedural knowledge is easily processed by computers. With remarkable recent advances in computing, we can now instantly carry out calculations that would previously have been impossible. As a result, it has become difficult to distinguish the technologies involved in these procedures, as they are imitated quickly. A company that first develops a procedure may achieve greater efficiency and be at an advantage for a short duration, but other companies soon catch up. When this happens, they must seek even greater efficiency in their development. The so called competition for the sake of competition will continue endlessly. Companies will be able to create products more cheaply and faster, but on the other hand, they will end up doing nothing else. In order to break free from this cycle as it were, of ants entrapped in a doodlebug's pit, companies must emphasize meaning in their designs. The inner motive in the designer's mind plays an important role in this. This is because it is the designer who can best understand the physical meaning of design and the social meaning of the products that are created as its result, given her/his earnest involvement. The designer must believe in her/his own inner motive and act boldly.

Meaning cannot be treated objectively. Nor is there true or false in meaning. So how should we treat meaning? Nobody has all the answers to this. Even so, the act of treating meaning in a certain manner at least means acknowledging that multiple meanings exist. In other words, designers are requested "to respect the meaning of the actions of others, and both boldly and humbly design while considering the meaning of their own actions." This could be described as the trait required of those who design technologies, and even of the discipline of engineering itself.

Task

In 2011, Japan experienced a once in 1000-year disaster, the Great East Japan Earthquake. A variety of research has been conducted on this, for example, an attempt was made to investigate the traces left by a tsunami that occurred 1000 years ago. However, not much is heard of how we can tell our descendants of this disaster on the huge time scale of 1000 years on. Why do you think such a situation exists?

REFERENCE

Yoshikawa, H., 1993. Techno-Globe. Kogyo Chosakai Publishing Co., Ltd, Tokyo, (in Japanese).

EPILOGUE

This book has outlined a theory and methodology of design. However, how can we obtain or develop such theory and methodology of design in the first place? In other words, how should design research be conducted? In fact, this question troubles many design researchers. I would like to discuss these aspects briefly to tie up the arguments in this book. First, I will look at what kind of question this is, by outlining some key aspects discussed in our previous work with some modifications (Perspectives on design creativity and innovation research, *International Journal of Design Creativity and Innovation*, vol. 1, no. 1, pp. 1–42 (2013)), although a narrative style is adopted in this discussion.

Indeed, design can be described using an analogy of "living things moving in a field." If you want to know about "living things moving in a field," there is no point in even trying unless one engages them while living and moving in the field themselves. This makes the process all the more difficult. One can photograph them, capture them on video, raise and breed them, cut them up and dissect them, or stuff and mount them, all in an attempt to get closer to them. Once one has done so, however, they are no longer really living things moving in a field, and they have been reduced to something completely different from what was once moving about vibrantly. In the end, such an approach cannot possibly tell a thing about the creatures living in the field.

The process known as research, particularly the writing of scientific papers, has its own limits in describing "living things" that are alive, because once one sets about the task of describing them, it is necessary to restrict or stop the movements of those "living things." Limiting their movements allows for more accurate descriptions, but their natural movements then slip away from the purview of such delineation. The attempt thus becomes futile, and one does not achieve the goal of comprehending their natural movements in the field successfully. What, therefore, can be done?

One possibility is to go out into the field and move with them. By moving with them, one might realize why they move and what it feels like to move like them. The drawback of this approach is that one cannot give an account of something while moving, and, therefore, it is necessary to remember what the movement felt like and articulate it later. That is the only way to go about it—one has to stop and record things before forgetting about them. Meanwhile, when moving, one must be completely absorbed and immersed in the movement. This is because there is no hope of coming close to the essence of the movement if one is simply acting impromptu without being engrossed in the act. We refer to this state of being absorbed and engrossed as "passion." However, one cannot provide an account with accuracy while being lost in the movement. To describe the movement accurately, a dispassionate view is needed. If only there were a way to provide an accurate account while retaining passion—unfortunately, such a way has yet to be discovered.

Another method is to raise living things well and observe them; however, this is easier said than done. To raise them well refers to raising them such that they begin to exhibit their wild instincts and run about in the field. The catch is that one cannot do this without a priori knowledge of how they actually live in the wilderness. Now, when it comes to design, what may seem to be taken for granted might not in fact be anything that should be viewed as only natural; rather, it may have actually been impossible from the start.

Design research approaches in a "laboratory setting" are similar to the latter method. To recapitulate what has been stated earlier, this is, however, definitely different from the conventional research approach in that one cannot conduct any experiments without knowledge of design. That is to say, a demonstrative experiment, rather than an experiment to gain knowledge, is required. Furthermore, because it must be an experiment to show the manner of existence of living beings, it must show the experiment participant animatedly running about the field; that is, the participant must be allowed to become passionate. Both the experimenter and the participant need to be enthusiastic and immersed. This could be achieved at the cost of the experiment's objectivity, as only the experimenter and the participant would know if they had become passionate and, if so, to what extent. Can we really expect to make accurate observations when the success or failure of an experiment is determined by such factors?

At the same time, as researchers, we cannot fulfill our roles if new findings are made known only to ourselves. Rather, it is important that they be "accumulated" for appreciation by future generations as part of an academic discipline. One might very well wonder what this academic discipline is. How should we understand the notion of "falsifiability," which is considered essential for scientific research?

In addition, the author is of the following opinion. When understanding objects or things, perspective plays a crucial role. What you can see or the "view" differs completely depending on what you are looking at (the focal point) and from where (the standpoint). Sometimes we engage in meaningless arguments over which is correct, even though it is obvious that the "view" will be different from different standpoints. The same sun at the same point in time can cause one person to be moved at a sunset and give another a sunrise to gaze at.

Our view of living designs also changes completely depending on which point we look at and from where. We may see them as solutions to problems, works of art, cognition processes, esthetic patterns, emergent phenomena, or as applications of physical phenomena. So, are such views only views after all, and is the essential nature of a design beyond our grasp? The author believes that if a variety of perspectives can be gathered and reconstructed, then the essential nature of a design will begin to emerge. It is like reconstructing a solid figure from a three-view drawing (front view, side view, and top view). Each view can only illustrate one perspective, but the collection of these different perspectives begins to give shape to the solid. If what we are observing is a living thing moving around, then we must observe it from an even greater number of angles.

What those who work with or conduct research on design ordinarily see is the view from only one perspective (angle) of design. Therefore, it is dangerous to define (prescribe) that limited view as design. Needless to say, it would be absurd to claim that one's view alone would help detail all the nuances that make up design. When integrating views from multiple perspectives, it is definitely important to clarify the standing position (perspective) of each. In this sense, it is meaningful to define design before stating one's view. The definition of design in this book is one such definition (the standing position is clearly stated).

This book itself is actually composed of several perspectives. Throughout my life, design has been the main theme. Although (in my opinion) the theme has remained the same, the perspectives I have directly and indirectly learned until now have been many and varied. In my specialist undergraduate training and postgraduate study at university, I studied Precision Engineering. After completing my postgraduate course, I was employed at Nippon Steel Corporation (now known as Nippon Steel & Sumitomo Metal), where I engaged in the practice of engineering for 9 years at Sakai Works. It was indeed a period when I was hounded by design and one in which I pursued the essence of design, passionately moving about within the "field." There, I learned the meaning behind designing machines from my seniors. In particular, I feel that Hideo Matsui imparted to me the attitude toward mechanical design that I hold today. For example, he once asked me the meaning of calculating the strength of a rolling mill, a large machine that is incomparably different from material specimens. The answer to such a question cannot be found in a material mechanics textbook.

After becoming a researcher, I was able to learn directly from Hiroyuki Yoshikawa. I feel that he taught me many things, such as what academics is, and, in particular, what engineering is, and about the relationship between academics and perspectives. Around 20 years ago, I was lucky enough to spend many hours discussing and debating with him in the office of the Dean of the Faculty of Engineering at the University of Tokyo. Those discussions about the essence of design are closely related to the principles of this book, and will forever stay with me. In addition, during the 6 years I spent working at Research into Artifacts, Center for Engineering (RACE) of the University of Tokyo, I was surrounded by passion, as it was a gathering of highly ambitious researchers who were aware of the need to build this new discipline. Interdisciplinary exchange between researchers in philosophy and product design also began in this period.

In the last 10 years, I have been blessed with a number of brilliant collaborative researchers, and I have been able to make concrete progress in integrating a range of the perspectives they have provided. Yukari Nagai has been kind enough to introduce to me the methodology of cognitive studies and how we view artistic objects, and from Eiko Yamamoto, I learned the methodology of natural language processing. Thanks to these two researchers, I have been able to passionately and falsifiably demonstrate the hypothesis that had been pondering based on my experiences and discussions. To do so, we started from scratch, developing a new methodology. For example, we developed a method in which experiment participants are videotaped while immersed in the act of designing, and are then shown these videos while asked to objectively confirm the reasons for their design activities. Furthermore, although the primary role of simulation is generally to recreate a phenomenon in the natural world within a computer, we have proposed a new idea called "constructive simulation," which uses simulation as a means for discovering principles that lie behind these phenomena. The former method is outlined in Section 8.4, while the latter is explained in Section 3.1. For several years now, I have been conducting multifaceted research on design with members of the Creative Design Laboratory at the Department of Mechanical Engineering, Kobe University—Akira Tsumaya and Kaori Yamada.

Meanwhile, I am also broaching this discussion on a global scale. In 2007, my colleagues and I formed an international community under the theme of "design creativity," one of the keywords of this book. These activities are developing each year, and comprise of holding regular international conferences and even publishing our own journal. These are not only forums for researchers to present their own research findings, but also outlets for us to work toward creating an all-encompassing overview of the relevant fields by providing numerous opportunities for each of us to discuss our own ideas and views. For example, our conference proceedings include not only papers presenting research findings but also position papers from researchers active in the cores of their own fields (*Design Creativity 2010*, Springer (2010)). The inaugural issue of our journal brings together perspectives about the relevant fields in the form of an extended editorial, and 36 globally active researchers have contributed their thoughts (Perspectives on design creativity and innovation research, *International Journal of Design Creativity and Innovation*, vol. 1, no. 1, pp. 1–42 (2013)). Such position papers and the extended editorial are read by a large number of researchers. Furthermore, in 2013, researchers from fields ranging from engineering design, industrial design, information science, architecture, cognitive science, art to even philosophy and ethics have come together to hold a workshop under the themes of "motive of design," and "pre-design phase, design phase, and post-design phase." The August heat in Nara in Japan could not match the heated discussions that took place. The participants have each passionately elucidated their thoughts based on the workshop discussions (*Principia Designae: Pre-Design, Design, and Post-Design: Social Motive for the Highly Advanced Technological Society*, Springer (2014)). Here, I would like to express sincere gratitude to Georgi V. Georgiev for his devoted contributions toward these international activities.

As mentioned previously, this book is the author's attempt at summarizing a theory and methodology of design, based on a variety of questions and answers that arose from collaboration with a range of people.

Finally, I would like to extend my heartfelt gratitude to all my seniors and colleagues thus far, and the students who have assisted with my research. I am also indebted to Rika Tannai, editor of the original publication of this book at the University of Tokyo Press. She has provided me with apt advice from the draft stages, which has greatly improved the content of this book. I also express my thanks to Elsevier Inc., especially to Jonathan Simpson, Brian Guerin and Carrie Bolger for publishing this book.

SUBJECT INDEX

Printed in the United States
By Bookmasters

Printed in the United States
By Bookmasters